CLIMATE CHANGE AND ITS CAUSES, EFFECTS AND PREDICTION

ASSESSING CLIMATE CHANGE IMPACTS, VULNERABILITY AND ADAPTATION: THE CASE OF PANTABANGAN-CARRANGLAN WATERSHED

CLIMATE CHANGE AND ITS CAUSES, EFFECTS AND PREDICTION

Additional books in this series can be found on Nova's website under the Series tab.

Additional E-books in this series can be found on Nova's website under the E-books tab.

CLIMATE CHANGE AND ITS CAUSES, EFFECTS AND PREDICTION

ASSESSING CLIMATE CHANGE IMPACTS, VULNERABILITY AND ADAPTATION: THE CASE OF PANTABANGAN-CARRANGLAN WATERSHED

RODEL D. LASCO
REX VICTOR O. CRUZ
JUAN M. PULHIN
AND
FLORENCIA B. PULHIN

Nova Science Publishers, Inc.
New York

For permission to use material from this book please contact us:
Telephone 631-231-7269; Fax 631-231-8175
Web Site: http://www.novapublishers.com

LIBRARY OF CONGRESS CATALOGING-IN-PUBLICATION DATA

Assessing climate change impacts, vulnerability and adaptation the case of pantabangan-carranglan watershed / Rodel D. Lasco ... [et al.].
p. cm.
Includes index.
ISBN 978-1-61122-132-9 (softcover)
1. Watersheds--Environmental aspects--Philippines--Case studies. 2. Water resources development--Environmental aspects--Philippines--Case studies. 3. Climatic changes--Philippines--Case studies. 4. Pantabangan River Watershed (Philippines)--Environmental conditions. I. Lasco, Rodel D.
GB992.P6A87 2010
551.4809599'1--dc22
2010035916

Published by Nova Science Publishers, Inc. † New York

CONTENTS

PREFACE

Climate change is one of the defining issues of our day. As I write this, world leaders are in Copenhagen grappling with how to deal with this seemingly intractable issue. While scientists are almost one in pointing out the dangers of climate change, global response is still fragmented.

The Philippines is one of the most vulnerable countries to climate risks both current and future. Annually, the country experiences the havoc caused by tropical cyclones. Every few years, the ENSO phenomenon adds another layer of confusion to the weather patterns. As a result, small farmers in the country have developed ways of adapting to such seemingly capricious climate.

This book arose out of pioneering research in one of the most important watersheds in the country. When it started in 2002, there was hardly any information on the impacts of climate change on Philippine watersheds and their natural and social systems. Through this research, we were able to explore how climate change could affect our forests, water resources and local communities.

The methodologies, key findings and lessons learned from this research are now compiled in this book. It is our hope that these be useful to researchers, students, policy makers and development workers interested on the impacts of climate change as well as how we can cope with it.

We would like to express our gratitude to the global project Assessment of Impacts and Adaptation to Climate Change (AIACC) of START and UNEP of which we had been a part. Special thanks go to Dr. Neal Leary who was the global coordinator of AIACC. But most of all, we are indebted to the small farmers and institutional partners in the Pantabangan-Carranglan Watershed who gave their time and resources in the conduct of this research.

LIST OF ACRONYMS / ABBREVIATIONS

A and D	alienable and disposable
AIACC	Assessments of Impacts and Adaptations to Climate Change
CAI	current annual increment
CLSU	Central Luzon State University
CLUE-S	Conversion of Land Use and its Effects at Small
CV and E	climate variability and extreme
DA	Department of Agriculture
DAR	Department of Agrarian Reform
Dbh	Diameter at breast height
DENR	Department of Environment and Natural Resources
DMPMC	Development Management Plan of the Municipality of Carranglan
ENSO	El Niño Southern Oscillation
FGD	focus group discussion
GCM	general circulation model
GIS	geographic information system
GHG	greenhouse gas
IA	irrigators' association
IEC	information, education and communication
IPCC	Intergovernmental Panel on Climate Change
LGU	local government unit
MAO	municipal agriculture officer
masl	meters above sea level
MAB	mean annual biotemperature
MAT	mean annual temperature
NDCC	National Disaster Coordinating Council
NIA	National Irrigation Authority
NIA-UPRIIS	Natioanal Irrigation Authority- Upper Pampanga River Integrated Irrigation System
NPC	National Power Corporation
NPP	net primary productivity
PAGASA	Philippine Atmospheric Geophysical and Astronomical Services Administration
PCW	Pantabangan-Carranglan Watershed

PET	potential evapotranspiration
PINC	Philippines Initial National Communication
PMP	Pantabangan Master Plan
PREC	precipitation in the growing season
ROC	Relative Operating Characteristic
SWIP	small water impounding projects/programs
TA	technical assistance
TAR	Third Assessment Report
UPRIIS	Upper Pampanga River Integrated Irrigation System
VIC	Variable Infiltration Capacity

Chapter 1

CLIMATE CHANGE AND WATERSHEDS

The Intergovernmental Panel on Climate Change (IPCC) Fourth Assessment Report concludes that climate change is becoming a present reality as warming of the climate system has become unequivocal (IPCC 2007). All general circulation models (GCMs) predict an enhanced hydrological cycle and an increase in area-averaged annual mean rainfall in Asia. This is expected to exacerbate pressure on the region's natural resources that are already under severe stress from rising population. Developing countries will be most vulnerable, as they have limited resources and capacity to adapt to the effects of climate change.

Apart from intraseasonal and interannual variability in climate, extreme weather events such as cyclones, prolonged dry spells, and intense rainfall are known to cause adverse effects such as droughts and floods in tropical Asia. Generally, what is observed in many parts of Asia is the increase in frequency of intense rainfall events that has caused severe flooding, landslides, and debris and mud flows (IPCC 2007).

Climate change poses varying threats, direct or indirect, both to natural and human systems. Scientific studies show that human health, ecosystems, and socioeconomic sectors (e.g., hydrology and water resources, food and fiber production, coastal systems and human settlements), all of which are vital to sustainable development, are sensitive to changes in climate. These will be affected by both the magnitude and rate of climate change, as well as the changes in climate variability (IPCC 2001 and 1996).

While many regions are likely to experience the adverse impacts of climate change, of which some are potentially irreversible, some effects of climate change are likely to be beneficial. For instance, crop production in the mid and high latitudes is projected to increase at a local mean temperature increase of 1-3 0C (IPCC 2007). Clearly, it is important to understand the nature of climate change risks, where natural and human ecosystems are likely to be most vulnerable, and what may be achieved by adaptive responses.

In Southeast Asia the key concerns include the impacts of climate change on ecosystem vulnerability (e.g., biodiversity loss) and water resources. The IPCC Third Assessment Report (TAR) highlights the scarce information available on these concerns. Specifically lacking are integrated assessments of impacts, adaptation, and vulnerability (IPCC 2001).

Watersheds are critical to economic development and environmental protection in Southeast Asia and are likely to be affected by future climate change. In the Philippines watershed areas are believed to be among those to be adversely affected by climate change. Watersheds are critical to the economic development and environmental protection and are, therefore, key to the pursuit of sustainable development. More than 70% of the country's total land area lie within watersheds. Much of the remaining natural forests that provide a host of environmental services are located in these areas. Also, it is estimated that no less than 1.5 million hectares of agricultural lands currently derive irrigation water from watersheds. Moreover, around 20 to 24 million people, nearly one-third of the country's total population, inhabit the uplands of many watersheds, majority of whom depend on watershed resources for survival. However, scientists working on climate change in Southeast Asia have limited experience in impacts and vulnerability assessment. Lack of research support from internal sources has stifled development of research capacity. Aside from resource constraints, strategic partnerships with scientists from developed countries are also required.

The lack of research is reflected in the absence of articles from the Southeast Asian region in peer-reviewed literature. The result is under-representation of cases from the region in the IPCC assessment reports.

The Assessments of Impacts and Adaptations to Climate Change (AIACC), a global study designed to address the lack of scientific research on climate change impacts, adaptation, and vulnerability of watershed resources and local communities, was launched in 2002. One of AIACC's regional studies focused on the Philippines, Indonesia and Indo-China. This project assessed the impacts of climate change and associated land use and cover change on water resources, forest ecosystems, and social systems of watersheds in Southeast Asia. The project leaders conducted studies in selected watersheds in the region and provided training and technical assistance to scientists from Indo-China on research methods to be implemented in their respective watersheds. Future climate scenarios were developed and downscaled, and the results were used in conjunction with a climate-vegetation model to predict future land use and cover change. The impacts of climate and land use/cover change were assessed with measures of change in biodiversity, carbon and water budgets, livelihood, health, demographic shifts, and changes in social structure resulting from climate and land use/cover change. The project team conducted an integrated vulnerability assessment of natural and social systems in the watershed. It also developed and evaluated adaptation strategies. Research findings and policy implications were presented to policy makers and development workers. In the Philippines the study was conducted in Pantabangan-Carranglan Watershed (PCW) in Nueva Ecijia province.

1.1. Overview of Pantabangan-Carranglan Watershed

1.1.1. Biophysical Environment

Geographic Location and Importance

The PCW lies between 15°44' to 16°88' north latitude and 120°36' to 122°00' east longitude (Figure 1.1).

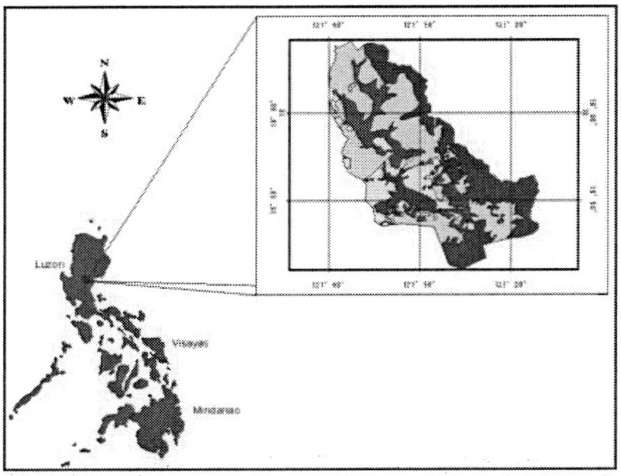

Sources: Landuse maps [1:50,000] from College of Forestry and Natural Resources, UPLB and Provincial Environment and Natural Resources Office, Nueva Ecija.

Figure 1.1. PCW location map.

The watershed is bounded on the north, northwest and northeast by the Caraballo Mountain Ranges and on the south, southeast and southwest by the Sierra Madre Ranges. It is located in the Municipalities of Pantabangan and Carranglan in the Province of Nueva Ecija, Municipalities of Alfonso Castañeda and Dupax del Sur in the Province of Nueva Vizcaya, and Municipality of Maria Aurora in the Province of Aurora. The watershed is approximately 176 km away from Manila (Saplaco et al. 2001).

PCW has a total area of 97,318 ha of which 4,023 ha comprise the water reservoir (1999 Land Use Map). It is considered as a 'critical watershed' under the government's classification since it supports a multi-purpose dam for irrigation and hydroelectric generation. The watershed also provides water for domestic and industrial uses and serves to tame the flood waters which for years damaged the farm crops in Central Luzon (NPC 1997).

The completion of the Pantabangan Dam in 1974 has greatly eased the water supply requirement of about 8,000 hectares of farmlands in the floodplains (NPC 1997). Currently, PCW supplies the irrigation requirements of 24 municipalities in Nueva Ecija, Bulacan, and Pampanga. It has a total service area of 102,532.21 ha which is divided into four districts. A total of 369 irrigators' associations (IAs) consisting of 62,039 farmers depend on PCW for their farm irrigation needs (NIA-UPRIIS 2004).

The construction of the Pantabangan Dam has also addressed the country's need for electricity. It generates 100,000 kilowatts of hydroelectric power which supplies electricity to Central Luzon and adjacent regions (NPC 1997).

Climate

PCW area largely falls under the Philippine Climatic Type I with two pronounced seasons, namely, dry from December to April and wet the rest of the year. A small portion of the watershed, especially that at the boundary of Aurora, falls under Climatic Type II, characterized by no dry season and very pronounced maximum rainfall from November to January. Its annual average rainfall is 1,766.5 mm (Saplaco et al. 2001, NPC 1995 and 1997).

Minimum monthly temperature is recorded at 23.21 0C and 33.71 0C for the maximum monthly temperature. The lowest temperature occurs during December through March while the highest temperature occurs from April through November with an average of 33 0C (NPC 1995 and 1997).

The average annual relative humidity is 83.37%. The lowest relative humidity occurs during May with 76.6% and the highest during September with 86.67% (NPC 1995 and 1997).

Topography and Soil

The topography of PCW is characterized by complex land configuration and mountainous rugged terrain. It ranges from nearly level, undulating and sloping to steep hilly landscapes. The highest mountain peaks in the area are Mount Susong Dalaga, with an elevation of 1,650 masl, and Mount Nedumular, with an elevation of 1,410 masl. Among the other mountains that can be seen in the area are Mount Amok, Kaanducian, Pulog Mabilog, Carranglan, and Maluyan. Strong, undulating, hilly terrain dissected by narrow, flat-bottom valleys formed by streams characterize the foot of these mountains. The headwaters of the streams originate from the mountains (Saplaco et al. 2001).

Soils at PCW originate mostly from weathered products of meta-volcanic activities and diorite. Surface soil textures are silty clay loam, clay loam, and clay. There are four types of soils in the watershed, namely: Annam, Bunga, Guimbaloan, and Mahipon (Saplaco et al. 2001). The Annam soil type is primarily a mountain soil which comes from weathered igneous rocks. It is moderately deep ranging from 50-130 cm and dominantly brown and clayey in color. This soil type is recommended for trees and forest crops. Guimbaloan soil is usually found on moderately sloping or undulating areas and on hilly and mountainous relief. Derived from basalt and meta-volcanic materials, this soil type is predominantly clayey, about 50 cm deep and well drained. The Bunga soil type, on the other hand, is present in level to nearly collu-alluvial landscapes. The dominant color is dark, grayish brown with strong brown and light gray matter. It is clayey in texture, with a depth of 147-155 cm and is moderately well drained. Mahipon usually occurs on level to nearly level collu-alluvial landscapes. This soil type is derived from quaternary alluvial/tallus deposits and terrace gravels. It also has a clayey texture but has a restricted internal drainage. It is moderately acidic which makes it useful for the cultivation of agricultural crops.

Land Use

The major land use types found in PCW are forests, open grasslands, and reforestation sites (Figure 1.2). Vegetation in the watershed is predominantly second growth. Since the logging boom in 1960s, primary forests in the watershed have greatly declined, though remnants of dipterocarp forest can still be found (Saplaco et al. 2001). Nevertheless, of significant occurrence is the increase in the area of reforested sites, although these sites are now under intense pressure from increasing population. Residential and barangay sites, as well as cultivated areas, are included in the alienable and disposable (A and D) areas.

Rice, vegetables, corn, onion, and other agricultural crops are grown on cultivated lands. Rice, onion, and vegetables are the primary crops raised on the lowland areas of Carranglan. Within the watershed divide, most of the areas devoted for rice production are rain-fed. Water pumped from wells and run-of-the-river irrigate some areas for rice production. With this current cultivation practice, rice growing cannot maximize the best use of the land. Hence, other primary crops are produced like yellow corn and onion. Secondary crops are also planted which include vegetables, like eggplant, tomatoes, bitter gourd, and squash (Saplaco et al. 2001, Toquero 2003).

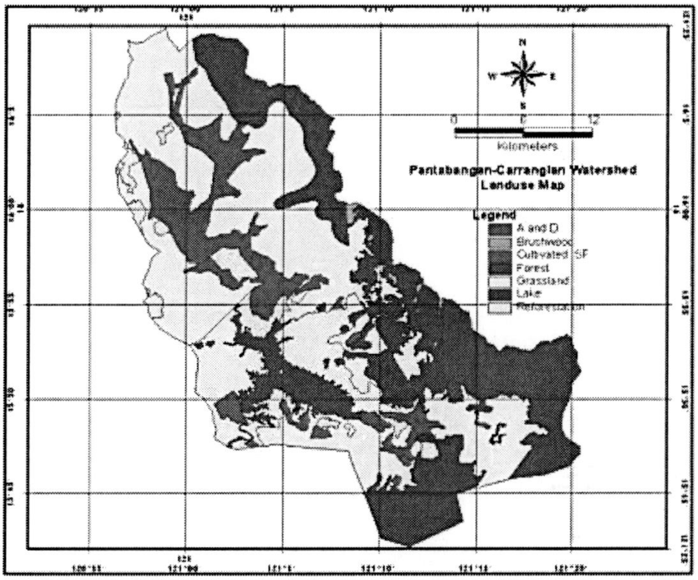

Figure 1.2. 1999 Land use map of PCW with municipal boundary.

The most popular crop during the dry season is onion. Second cropping is practiced by farmers in irrigated rice lands and sometimes in rain-fed areas depending on the frequency of rain during the year (Toquero 2003). In kaingin (slash-and-burn) farms crops like banana, cassava, sweet potatoes, and corn are normally grown. Meanwhile, only one pastureland was recorded in the area. This is found in Pantabangan where only limited areas are devoted for agriculture (Saplaco et al. 2001).

1.1.2. Demographic Characteristics

There are a total of 36 barangays found in PCW, of which 17 are found in Carranglan, 14 in Pantabangan, three in Alfonso Castañeda, and two in Ma. Aurora. As of year 2000 about 60,824 people reside in PCW which comprise 12,398 households (Census 2000).

Three ethnic groups inhabited PCW long before the Spanish occupation. These are the Aetas, Irol-les and the Italengs. They were soon joined by several groups of migrants, among them were Pangasinensis, Ibaloi, Ifugao, Waray, Bicolano, Pampango, Kalinga, Kankanai, Ibanag, Cebuano, and Ilongot. However, the construction of the Pantabangan Dam in 3 March 1971 has led to relocation of the residents of the town which resulted to waves of out-migration from the period of 1970s to 1980s. Today, residents in PCW are predominantly Tagalog and Ilocano. Other groups present in the area are Pangasinensis, Pampango, Waray, Bicol, Ifugao, and Ibaloi (Saplaco et al. 2001).

1.1.3. Socioeconomic Characteristics

The largest portion of PCW is located in Pantabangan and Carranglan in Nueva Ecija. The major source of livelihood of these municipalities comes from agricultural activities. In Pantabangan 13% of the total land area which accounts for about 5,406 ha are devoted to agriculture. Meanwhile, a total of 19,704 ha or 28% is allotted for farming in Carranglan. Among the major crops produced are rice, corn, onion, and vegetables. However, even if the Pantabangan reservoir is located in these areas, it only acts as a host for irrigation water to the Central Luzon area. Farmlands are unirrigated because of topography, hence, farmers are dependent on rain (Master Plan of the Municipality of Pantabangan [MPMP] 1998-2000, Development Management Plan of the Municipality of Carranglan [DMPMC] 2003-2007).

Fishing is the second largest industry in these areas, specifically in Pantabangan. This is because the area houses the dam reservoir which is one of the biggest fishing reservoirs in Asia. Carranglan, on the other hand, depends on large fishponds for its fish production. Other sources of income of residents are cottage and business activities which include wood and rattan craft, animal dispersal, and small stores (MPMP 1998-2000, DMPMC 2003-2007).

More than half of the productive population of Pantabangan and Carranglan are in the labor force. However, unemployment is still a problem due to limited employment opportunities in these areas (MPMP 1998-2000, DMPMC 2003-2007). Hence, many residents depend on the goods and services provided by the watershed for their livelihood. Commonly practiced in these areas are kaingin and charcoal-making.

Construction of Pantabangan Dam begun on March 1971 and was completed on August 1974. The man-made lake that forms part of the dam reservoir has submerged the old Pantabangan town and seven outlying barrios, namely: Liberty, San Juan, Cadaclan, Napon-Napon, Marikit, Villarica, and Conversion (Saplaco et al. 2001). All the residents of the old town were resettled to the upper portion of Pantabangan. This resettlement process, which was a joint responsibility of the National Irrigation Administration (NIA) and Department of Agrarian Reform (DAR), started in May 1973 and was completed in August 1974. NIA was in charge of most of the infrastructure development. DAR took care of land distribution for the settlers (Toquero 2003). The families affected received land grants in place of their submerged lands and were subsidized by the government for five years to include free water supply and lower electricity rate. Since the submergence of the town, the government has poured in livelihood projects to help residents recover from the loss of their properties and to make the place productive (MPMP 1998-2000).

Since the dam's construction, the area has continually received support from various agencies and institutions in the form of projects or programs. Uplifting the economic conditions of the relocated settlers was a prime concern of the government. DAR was the leading agency that took care of this mission. Even before the resettlement of the affected families, the agency started to conduct trainings on livelihood programs. This was followed by livestock dispersal (pig, goats, and cattle) and fish production projects. The Department of Agriculture (DA) took over the enhancement of the economic welfare of the residents in the mid-1980s. Livestock production was improved by the dispersal of bull and breeders in 1985-1998, and cattle carabao, goat, swine, and chicken in 1993-1998. Rice and corn production was also boosted in the same period through the distribution of certified seeds, organic fertilizers, use of small farm machineries, and a limited number of post harvest facilities (Toquero 2003).

One of the most prominent projects implemented in the watershed was the RP-Japan Reforestation Project which was launched in partnership with the Department of Environment and Natural Resources (DENR). This Japan International Cooperation Agency-funded project commenced in 1976 and ended in 1992. It aimed to reforest the open and denuded areas of PCW and provide technical support through the establishment of the Afforestation Technical Cooperation Center and the Training Center for Forest Conservation. The project has not only rehabilitated the denuded parts of the watershed but also created jobs for the local residents. Moreover, more than 600 Filipino forestry personnel were trained through this project who are now actively working in environment departments (Yoshida 2000).

Aside from the joint project with the Japan government, the DENR launched several reforestation programs, particularly in Pantabangan. These are the Regular Reforestation Program covering a total area of 823 ha and the Integrated Social Forestry Program which reforested 856 ha. The department has also engaged in Contract Reforestation Program with NIA in 1989-1990. In this program the DENR contracted NIA to reforest a total of 900 ha in PCW (MPMP 1998-2000).

The National Power Corporation (NPC) and NIA also have their share of projects implemented in the watershed area. Aside from training and extension services, NPC conducts yearly reforestation and extension projects in the three sectors under its jurisdiction. The reforestation projects cover an average of 30-40 ha a year.

Meanwhile, the biggest project implemented by NIA in PCW is the Watershed Management and Erosion Control Project which lasted from 1980–1988. This project was funded by World Bank and aimed to control soil erosion and minimize sedimentation and siltation in the reservoir. It has four components: reforestation, feasibility study of an integrated development, waste management and smallholder agroforestry pilot project, and integrated forest protection pilot program. Among the activities undertaken in this project were reforestation with agroforestry and timber crops of a 24,522-ha area in the watershed and construction of several plantation access roads and facilities. Due to the project's labor-intensive activities such as nursery operations, plantation establishment and development, and forest protection and fire prevention, it has provided employment for some 3,800 residents in Pantabangan in 1982. Aside from this, the project also provided revenue and profit share to the communities in the watershed in the form of facilities, such as domestic water supply, school building, and road improvements (NIA n.d.).

The Casecnan Multipurpose and Irrigation Project is a recent project implemented in PCW. It was constructed in November 1995 and began its operation on 11 December 2001. The project was designed to collect a portion of the waters of the Casecnan and Taan Rivers in Nueva Vizcaya and transport it to the Pantabangan reservoir. It is designed to benefit the irrigation requirement of 35,000 new hectares of agricultural lands and stabilize the water supply of the current areas serviced by PCW. Moreover, it will generate approximately 150 megawatts of hydroelectric capacity to the important Luzon grid.

As already mentioned, the above projects have significantly helped the residents in PCW through the provision of jobs, livelihood programs, and various forms of assistance. But despite the three-decade development effort of the government (amounting to PHP 1.5 billion), there is still widespread poverty in the resettlement as shown by a high percentage of families with income below the poverty threshold of PHP 7,377. The residents also perceived the services provided by the government organizations as unsatisfactory. This implies the failure of the government in providing an economically-viable resettlement area for the residents. A point of concern that could have contributed to this failure is the lack of participation of the residents in the planning and monitoring of the development projects or programs. Some residents were not used to the livelihood activities introduced, hence, they were forced to open kaingin in the critical watershed (Toquero 2003).

Moreover, these development projects and programs may have also resulted to dependency of some people to these forms of assistance and even to the goods and services provided by the watershed for their source of living.

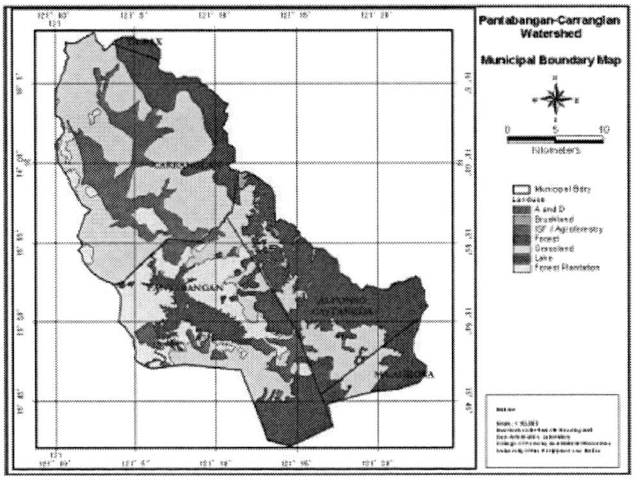

Figure 1.3. Land use map of PCW with municipal boundary.

With the recent completion of these development projects and programs, the local settlers resort to charcoal-making which destroys the areas that they reforested. What aggravates the situation is that this type of livelihood is practiced by more than 50% of the residents in the watershed (Toquero pers. comm.)

Irrigation Service

The PCW supplies the irrigation requirements of 24 municipalities in Nueva Ecija, Bulacan, and Pampanga (Figure 1.3) through the Upper Pampanga River Integrated Irrigation System (UPRIIS) which is operated by NIA. As shown in Table 1.1, it has a total service area of 102,532.21 ha which is divided into four districts. A total of 369 IA's consisting of 62,039 farmers depend on PCW for their farm irrigation needs (NIA-UPRIIS 2004).

Spearheading the management of PCW are national government agencies, namely, the DENR, NIA, and NPC. Each institution has specific areas within the watershed that is under its jurisdiction. This institutional arrangement comes from the need to sustainably manage the watershed so that there will be sufficient amount of water in the reservoir for irrigation and hydroelectric power generation. Supporting these institutions in the performance of their functions are the local government units (LGUs) present in the area.

The DENR is the primary government agency responsible for the conservation, management, development, and proper use of the country's environment and natural resources, including those in reservations, watershed areas, and lands of the public domain. It is also responsible for the licensing and regulation of all natural resources utilization as may be provided by law in order to ensure equitable sharing of the benefits derived for the welfare of the present and future generations of Filipinos. Since PCW is located in Central Luzon, it falls under the jurisdiction of DENR Region III. Two Provincial Environment and Natural Resources Offices and Community Environment and Natural Resources Offices located in Aurora and Nueva Ecija manage the watershed.

NIA is a government-owned and controlled corporation tasked with the development and operation of irrigation systems all over the country. Through Letter of Instruction 1002 dated 20 March 1980, NIA was authorized to manage, protect, develop, and rehabilitate certain portions of PCW Reservation. Meanwhile, Republic Act 3601, as amended by Presidential Decrees 552 and 1072, empowered NIA to acquire real and personal properties and all appurtenant rights, easements, concessions and privileges, whether the same are already devoted to private or public use in connection with the development of its projects (Saplaco et al. 2001). Under the jurisdiction of NIA are the areas of Ma. Aurora in Aurora, Alfonso-Castañeda in Nueva Vizcaya, and Carranglan and Pantabangan in Nueva Ecija (NIA 2003).

Meanwhile, NPC is the agency engaged in power generation and transmission all over the country. Among its functions is the development of electric power generation facilities including hydroelectric power and construction and operation of dams, reservoirs, and diversion facilities required for this purpose.

Table 1.1. PCW service area

Province / Municipalities	District 1	District 2	District 3	District 4
NUEVA ECIJIA				
1. San Jose	5,727.72			
2. Muñoz	4,032.62			
3. Sto.Domingo	5,547.04			
4. Quezon	3,670.56			
5. Licab	2,404.67			
6. Llanera	1,140.39	4,899.85		
7. Talavera	2,439.00	5,751.79		
8. Rizal		4,664.24		
9. Gen. Natividad		6,608.75	254.75	
10. Aliaga		1,686.12	2,969.45	
11. Cabanatuan City		302.17	6,002.13	
12. Sta. Rosa			5,481.45	
13. San Leonardo			2,893.04	
14. Jaen			5,069.98	
15. Zaragosa			2,404.00	
16. San Antonio			4,437.31	
17. Peñaranda			334.18	379.00
18. Gapan				5,259.63
19. San Isidro				3,448.82
20. Cabiao				
BULACAN				
1. San Miguel				4856.85
2. San Ildefonso				517.88
PAMPANGA				
1. Arayat				1462.78
2. Candaba				3312.13
Sub-Total	24,962.00	23,912.92	29,846.29	23,811.0
GRAND TOTAL				102,532.21

A memorandum of agreement signed on 1 April 1997 turned over to NPC a total of 14,166 hectares of PCW. The areas under the jurisdiction of NPC are Daldalayap, Bunga, and

Carranglan of Sector I; Burgos, Carranglan of Sector II; and Conversion, Pantabangan of Sector III (NPC 1997).

Finally, the LGUs constitute the provincial, municipal, and barangay political units in the Philippines. They can engage in development programs, economic activities, law enforcement, and legislation, among others. Through the process of devolution instituted under the 1991 Local Government Code, the LGUs are given the responsibility to conserve, manage and protect the natural resources.

Chapter 2

METHODOLOGIES

This chapter presents the methods and simulation models used in the AIACC study. This will be useful to students and other researchers who want to engage in climate change research. Further details of the methods and models are given in the references.

Climate change impacts on Philippines forest ecosystems are simulated using Holdridge Life Zones and geographic information system (GIS). The CO2Fix Model is used to quantify carbon stocks and fluxes of various land cover types in PCW. The CLUE-S (Conversion of Land Use and its Effects at Small) Model is used to predict land use change and land use in PCW. Impact assessment on water resources is done using SEA/BASINS Integrated Modeling System and the BROOKS Hydrologic Models. Meanwhile, the socioeconomic component of the study uses a combination of qualitative and quantitative methods in order to arrive at a multi-level indicator of vulnerability of communities living in PCW.

2.1. SIMULATING CLIMATE CHANGE IMPACTS ON FOREST ECOSYSTEMS USING HOLDRIDGE LIFE ZONES AND GIS

Generally, the study follows the approach of using scenarios to predict forest cover that is used by Somaratne and Dhanapala (1996) in Sri Lanka and is given as an example in the Adaptation Policy Framework of the United Nations Development Programme (Jones and Boer 2003).

The Holdridge Life Zone system is an ecological classification system based on three climatic factors, i.e. precipitation, heat (biotemperature) and humidity (potential evapotranspiration ratio (Holdridge 1967). Holdridge (1967) defined a life zone as a group of associations related through the effects of these three major climatic factors. Figure 2.1 shows the most common life zones on the Earth. All Philippine forests can be classified under the tropical belt because biotemperature is always greater than 24 0C. Thus, the main determinant of life zone classification is precipitation. The value used for precipitation is the mean annual total of water in millimeters that falls from the atmosphere either as rain, snow, hail, or sleet. In the case of the Philippines, this is the mean annual rainfall.

Precipitation = annual rainfall (mm)

The mean annual biotemperature is the measure of heat that is utilized in the life zone chart. The biotemperature mean is the average of the Celsius temperatures at which vegetative growth takes place relative to the annual period. The range of temperatures within which vegetative growth occurs is estimated to lie between 0 0C as a minimum and 30 0C as a maximum. The positive temperatures within this range must be averaged out over the whole year period in order to make it possible to effectively compare a given site with any other on the earth. The equation for biotemperature is as follows:

Mean annual biotemperature (MAB) = Σ (0 < T < 30)/ 12 months

Or

Mean annual biotemperature = Σ (0 < T < 30)/365 days

The third climatic factor that determines the boundaries of life zones is humidity, best described by the potential evapotranspiration (PET) ratio. PET is the theoretical quantity of water that would be given up to the atmosphere within a zonal climate and upon a zonal soil by the natural vegetation of the area throughout the growing season. Since both evaporation and transpiration are directly correlated with temperature, other factors being equal, the mean annual potential evapotranspiration in mm at any site may be determined by multiplying the mean annual biotemperature by the factor 58.93.

The PET ratio is determined by dividing the value of the mean annual potential evapotranspiration by the value of the mean annual precipitation, both in millimeters. Since potential evapotranspiration is the total water that could potentially be utilized by the normal quantity of water available for potential use in transpiration by the vegetation or evaporation, potential evapotranspiration ratio is a measure of humidity that may be utilized for comparing distinct sites. In equation:

Mean Annual PET= (MAB) (58.93)

PET ratio= Mean Annual PET/Mean Annual P

The Holdridge Life Zone chart (Figure 2.1) is a graphical classification of some of the most common lifezones on earth for the purpose of showing relationships of mountain vegetation to that of lowlands, working only with annual precipitation and temperature values to develop this chart (Holdridge 1967). The Holdridge forest types utilized in this model are rough estimates of the potential forest types that will thrive for a given precipitation and temperature.

Based on the three parameters, Holdridge life zones for the Philippines are identified using ArcView 3.2. Changes in the distribution of forest types in response to synthetic climate change scenarios are determined (Table 2.1). These precipitation and temperature scenarios are within the limits of GCM projections in the country (Philippines Initial National

Communication [PINC] 1999). ArcGIS 8.1 is used to process the maps needed for the Holdridge life zone model.

Rainfall map is based on the data collected by the Philippine Atmospheric, Geophysical, and Astronomical Services Administration (PAGASA). Average rainfall (1961-1990) in the Philippines ranges from 1000–4000 mm. Temperature data is also gathered from PAGASA. A Thiessen map is created from the 55 stations all over the Philippines. Average temperature (1949-2002) in the country ranges from 19.3–28.2 0C.

Table 2.1. Synthetic climate change scenarios used in the study

Increase in rainfall (% relative to present)	Increase in temperature (0C)		
	1	1.5	2.0
25	Scenario 1a	Scenario 1b	Scenario 1c
50	Scenario 2a	Scenario 2b	Scenario 2c
100	Scenario 3a	Scenario 3b	Scenario 3c

The land use map is based on a 1993 map prepared by the Presidential Task Force on Water Resources Development and Management. There are only about six million hectares of forests left (excluding brushland and man-made forest), a mere 20% of the country's total land use. Of these, 1.6 million hectares are non-production forests and less than 1 million hectares are old growth forests. This land use map served as the boundary of Philippine forests and used as an overlay on the calculated Holdridge life zone in the Thiessen map.

2.2. Forest Carbon Budget

2.2.1. Field Sampling, Biomass Calculation, and Carbon Analysis

For natural forests the point-centered quarter method is used. It is a plotless method of sampling that is designed to determine the number of trees per unit area that can be calculated from the average distance between the trees. Four parallel lines are randomly laid out. Each parallel line comprises five sampling points with a 50 meter distance from each other. In each point two lines form a cross. One of the lines is the compass direction and the other is the line perpendicular to it and passes through the sampling point. The distance to the midpoint of the nearest tree inside each quadrat with a diameter at breast height (dbh) of ≥ 10 cm is measured. The tree species is then identified and its dbh measured.

For brushlands, 10 plots measuring 10 m x 10 m are established. In each plot all trees with > 10 cm dbh are identified and their dbh measured. For tree plantations, two 5 m x 40 m plots are established in each plantation type. If trees with dbh > 50 cm are present whether they are included in the sample plots or not, an additional sample of 20 m x 100 m is established. In plots measuring 5 m x 40 m, all trees with dbh of > 5 cm are measured and identified. In 20 m x 100 m plots only trees with a diameter > 30 cm are measured. For understorey vegetation, four frames measuring 1 m x 1 m are randomly laid out near the sampling points. All herbaceous and woody vegetation (< 5 cm dbh) inside the frame are collected. Sample fresh weight is determined and then samples are oven dried. Inside the same sampling frames used for measuring understorey and herbaceous vegetation, a 0.5 m x 0.5 m transect is established for litter collection. Total fresh weight of all the samples are taken after which about 300 grams are reserved for air drying and oven drying. Samples are dried inside the oven with a temperature of ± 102 °C for at least 48 hours or until weights of the samples become constant. Coarse litter is collected in the 0.5 m x 0.5 m quadrat within the understorey sample plot. Similar to the understorey, a sub-sample of about 300 g is taken for oven drying and carbon content analysis.

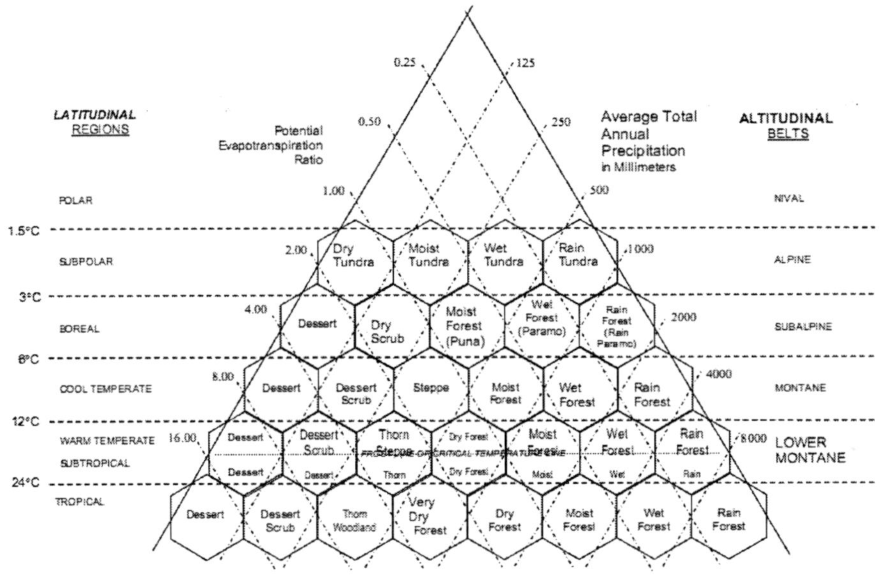

Figure 2.1. The Holdridge System of vegetative cover classification (Holdridge 1967).

In grasslands ten 1 m x 1 m sampling frames are laid on the ground. Grasses inside the sampling frames are harvested for biomass determination. Similar to understorey and litter, fresh samples are weighed and a sample of 300 grams is set aside for oven drying.

Soil samples are collected within the sample plots of second growth brushland and grassland areas at 30 cm depth. Samples are air dried and are taken to the Soils Laboratory of the Soil Science Department of the College of Agriculture, University of the Philippines Los Baños for analysis. For bulk density determination, samples are collected using a ring metal with height of 10 cm and diameter of 3 cm at 20-30 cm depth.

Tree biomass is calculated using the following allometric equation (Brown 1997):

Y (kg) = EXP(-2.134+2.53*LN(D))

Biomass values for litter, understorey and grasses are calculated using the following formula:

$$ODW_t = \frac{TFW - (TFW*(SFW-SODW))}{SFW}$$

where:

ODW = total oven dry weight

TFW = total fresh weight

SFW = sample fresh weight

SODW = sample oven-dry weight

Carbon density is calculated using the following formula:

C density (Mg ha-1)= Biomass density (Mg ha-1) * 0.45

For more details of field sampling of carbon stocks and sequestration, please refer to IPCC GPG (2003), Pearson et al. (2005), and the 2006 IPCC GHG inventory guidelines.

2.2.2. The CO2Fix Model

Total carbon budget of the various land cover in PCW is computed using the CO2Fix model. It is a tool which quantifies the C stocks and fluxes in the forest, soil organic matter compartment and the resulting wood products at the hectare scale. Version 2.0 of the model includes the following features:

- Simulate multi-species and unevenly-aged stands in multiple cohorts (defined as group of individual trees or group of species which are assumed to exhibit similar growth and which maybe treated as single entities within the model);
- Parameterize the growth by stand density;
- Deal with intercohort competition;
- Allocation, processing lines, and end-of-life disposal of harvested wood;
- Soil dynamics;
- Handle a wider variety of forest types including agroforestry systems, selective logging systems and post harvesting mortality; and
- Output viewing charts.

In modeling the growth of the stand, two basic approaches are considered:

- tree growth as a function of tree or stand age; and
- tree growth as a function of tree size or stand basal area, volume, or biomass.

Biomass Components

The model is parameterized to PCW conditions. Where the age of the stand is known, biomass growth is in the form of current annual increment (CAI) of stemwood volume, in m3 ha-1 yr-1. Moreover, biomass of other stand components such as branches, foliage, and roots are also calculated as an additional fraction to the growth rate of the tree biomass. Carbon content of the wood components and wood density of the sample species are also important inputs.

Turnover Rate

Data on the annual rate of mortality (turnover) of the biomass components is estimated using the default value while the stem turnover rate is a function of the mortality rate of the stand. The mortality of each species is described as the mortality due to senescence or old age and density-related competition and the mortality caused during and after logging operations. It is assumed that all trees have a maximum age and the annual mortality increases when the age of the stand approaches the maximum age, thus, in this model mortality is parameterized as a percentage of the standing biomass.

Interactions (Competitions)

Growth of trees in a stand is affected by the presence of other trees. There are two ways of parameterizing competition:

- Competition relative to the total biomass in the stand (e.g., in a Eucalyptus stand only); and
- Competition relative to each cohort (e.g., Eucalyptus and Acacia plantation).

A default value of '1' is used for no competition at all.

Management Interventions (Harvesting)

The inputs necessary to determine the management interventions are: the age at which the harvesting takes place, intensity of harvesting done (fraction of cohort biomass removed), and allocation of the biomass removed to different 'raw material' classes as slash, logwood, and pulpwood.

Parameterizing The Soil

The model requires inputs like the mean annual temperature (MAT) of the soil (0C), precipitation in the growing season (PREC, mm) and PET in the growing season (PET, mm) for the study site. PREC and MAT data can be found at http://www.worldclimate.com or mean monthly temperatures can be computed using 'PET.xls' file installed in CO2FIX. Values of litter in the soil are also needed.

Parameterizing Wood Products

The end of this module is to track carbon from harvesting to final decay. This is because carbon is released to the atmosphere when either by-products are set aside in the manufacturing phase, firewood is burned, or products in landfills decompose. The model allocates harvested biomass like the fraction of the logwood to be used for sawn wood, boards and panels, pulp and paper, and firewood. The fraction of pulpwood to be used for boards and panels, pulp and paper, and firewood are also allocated.

Information on processing losses is also necessary like the fraction of sawn wood lost and reprocessed as boards, pulp and paper, wood fuel, or wood dumped at the mill site.

Moreover, in the end-products stage the allocation of the commodities to different categories will be specified into long-term, medium and short-term use. The life span of the products in use describe the share of remaining products disposed each year.

Sources of errors in using the model are measurement errors in the data and the lack of data used for model construction. This study tries to collect all the necessary inputs needed by the model in the field in order to arrive at accurate projections and results. However, in nature variability occurs such as growth variation between years due to weather circumstances, intraspecies genetic differences and site quality variation and other management irregularities and risks caused by storm and fire. These variations are not captured by the model because it relies on fixed input data from perfectly managed stands.

2.3. Land Use and Cover Change: CLUE-S Model

This section describes in general the CLUE-S Model for land use and land cover change based on Verburg (2002). Details of how the model is used in PCW are described in Chapter 6.0.

Model Structure

The model is subdivided into two distinct modules: a non-spatial demand module and a spatially explicit allocation procedure (Figure 2.2). The non-spatial module calculates the area change for all land use types at the aggregate level. Within the second part of the model these demands are translated into land use changes at different locations within the study region using a raster-based system.

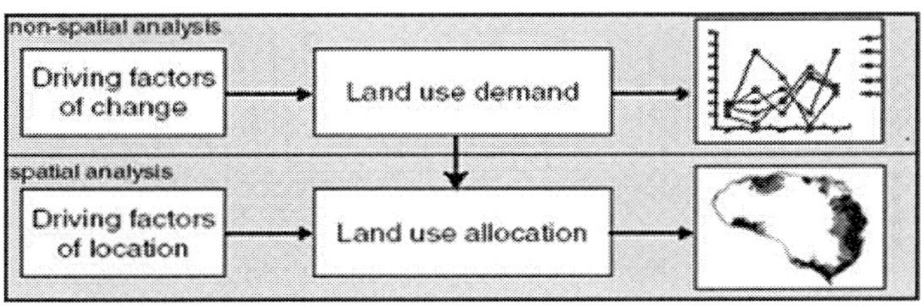

Figure 2.2. Overview of the CLUE-S modeling process.

For the land use demand module different alternative model specifications are possible ranging from simple trend extrapolations to complex economic models. The choice for a specific model is very much dependent on the nature of the most important land use conversions taking place within the study area and the scenarios that need to be considered. The results from the demand module need to specify, on a yearly basis, the area covered by the different land use types, which is a direct input for the allocation module. The rest of this discussion focuses on the procedure to allocate these demands to land use conversions at specific locations within the study area.

The allocation is based upon a combination of empirical, spatial analysis, and dynamic modeling. Figure 2.3 gives an overview of the procedure. The relations between driving forces and the spatial distribution of land use are determined by an empirical analysis of actual land use patterns. The results of this empirical analysis are used within the model when simulating the competition between land use types for a specific location. In addition, a set of decision rules is specified by the user to restrict the conversions that can actually take place based on the actual land use pattern. The different components of the procedure are now discussed in more detail.

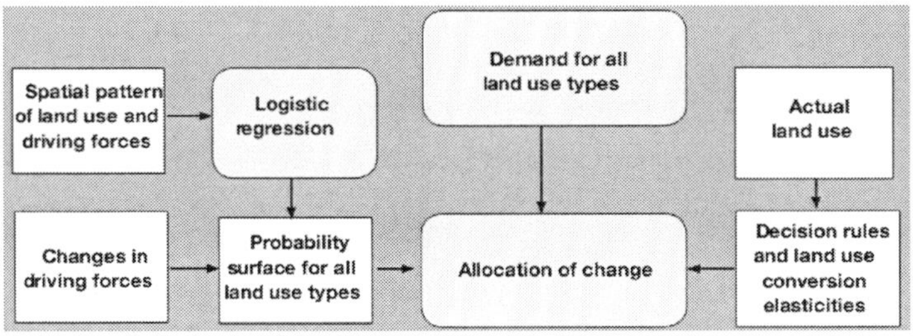

Figure 2.3. Schematic representation of the procedure to allocate changes in land use to a raster based map.

Spatial Analysis

The pattern of land use, as it can be observed from an aeroplane window or through remotely sensed images, reveals the spatial organization of land use in relation to the underlying biophysical and socioeconomic conditions. These observations can be formalized by overlaying this land use pattern with maps depicting the variability in biophysical and socioeconomic conditions. GIS is used to process all spatial data and convert these into a regular grid. Apart from land use, data are gathered that represent the assumed driving forces of land use in the study area. Data can originate from remote sensing (e.g., land use), secondary statistics (e.g., population distribution), maps (e.g., soil) and other sources. To allow a straightforward analysis, the data are converted into a grid-based system with a resolution that depends on the resolution of the available data. This often involves the aggregation of one or more of the thematic data. For instance, it does not make sense to use a 30 meter resolution if that is available for land use data only, while the digital elevation model has a resolution of 500 meter. Therefore, all data are aggregated to the same resolution that best represents the quality and resolution of the data. The relations between land use and its driving factors are thereafter evaluated using stepwise logistic regression. Logistic regression is an often used methodology in land use change research (Geoghegan et al. 2001, Serneels and Lambin 2001). In this study logistic regression is used to indicate the probability of a certain grid cell to be devoted to a land use type given a set of driving factors following the equation:

$$\text{Log}(\tilde{P}_1 Pi) = \tilde{\beta}\beta_1 X_{1i}\beta 2X2i + \beta n X n i$$

where Pi is the probability of a grid cell for the occurrence of the considered land use type and the Xs are the driving factors. The stepwise procedure is used to help select the relevant driving factors from a larger set of factors that is assumed to influence the land use pattern.

Variables that have no significant contribution to the explanation of the land use pattern are excluded from the final regression equation. The goodness of fit can be evaluated for all equations with the Relative Operating Characteristic (ROC) Method (Pontius and Schneider 2000, Swets 1986) which evaluates the predicted probabilities over the whole domain instead of only evaluating the percentage of correctly classified observations at a fixed cut-off value. This is an appropriate methodology for the application. A wide range of probabilities will be used within the model calculations which only makes the quality of the regression model over a wide range more interesting. The influence of spatial autocorrelation on the regression results can be minimized by performing the regression only on a sample of pixels that are selected at more or less equidistant from one another. Such a selection method is adopted in order to maximize the distance between the selected pixels to attenuate the problem associated with spatial autocorrelation.

Based upon the regression results a probability map can be calculated for each land use type. A new probability map is calculated every year with updated values for the driving factors that are projected to change in time, for instance, the population distribution.

Decision Rules

Land use type or location specific decision rules can be specified by the user. Location specific decision rules include the delineation of protected areas such as nature reserves. If a protected area is specified, then no changes are allowed within this area. Land use type specific decision rules determine the conditions under which a land use type is allowed to change in the next time step. These decision rules are implemented to give certain land use types a certain resistance to change in order to generate the stability in the land use structure that is typical for many landscapes. Three different situations can be distinguished and for each land use type the user should specify which situation is most relevant for that land use type.

For some land use types it is very unlikely that they are converted into another land use type after their first conversion. For instance, as soon as an agricultural area is urbanized it is not expected to return to agriculture or to be converted into forest cover. Unless a decrease in area demand for this land use type occurs the areas covered by this land use are no longer evaluated for potential land use changes. If this situation is selected it also holds that if the demand for this land use type decreases, there is no possibility for expansion in other areas. In other words, when this setting is applied to forest cover and deforestation needs to be allocated, it is impossible to reforest other areas at the same time. Other land use types are converted more easily. A swidden agriculture system is most likely to be converted into another land use type soon after its initial conversion. When this situation is selected for a land use type, no restrictions to change are considered in the allocation module. There are also a number of land use types that operate in between these two extremes. Permanent agriculture and plantations request an investment for their establishment. It is, therefore, not very likely that they are converted very soon after into another land use type. However, in the end, when another land use type becomes more profitable it is well possible that there is a conversion. This situation is dealt with by defining relative elasticities for change (ELASu) for the land use type considered ranging between 0 (similar to situation 2) and 1 (similar to situation 1). The higher the defined elasticity, the more difficult it gets to convert this land use type. The elasticity should be defined based on the user's knowledge of the situation, but can also be tuned during the calibration of the model.

Competition and Actual Allocation of Change

Given the probability maps, the decision rules in combination with the actual land use map and the demand for the different land use types, the actual allocation of land use change is made in an iterative procedure (Figure 2.4). The following steps are followed in the calculation:

1. The first step includes the determination of all grid cells that are allowed to change. Grid cells that are either part of a protected area or under a land use type that is not allowed to change (situation 1 above) are excluded from further calculation.
2. For each grid cell i the total probability (TPROPi,u) is calculated for each of the land use types u according to: u u u i u i ITER ELAS P TPROP, where ITERu is an iteration variable that is specific to the land use. ELASu is the relative elasticity for change specified in the decision rules (situation 3 described above) and is only given a value if grid-cell i is already under land use type u in the year considered. ELASu equals zero if all changes are allowed (situation 2).
3. A preliminary allocation is made with an equal value of the iteration variable (ITERu) for all land use types by allocating the land use types with the highest total probability for the considered grid cell. This will cause a change in land use for a number of grid cells.
4. The total allocated area for each land use is now compared to the demand. For land use types where the allocated area is smaller than the demanded area the value of the iteration variable is increased. The value is decreased for land use types for which too much is allocated.
5. Steps 2 to 4 are repeated as long as the demands are not correctly allocated. When allocation equals demand the final map is saved and the calculations can continue for the next yearly timestep.

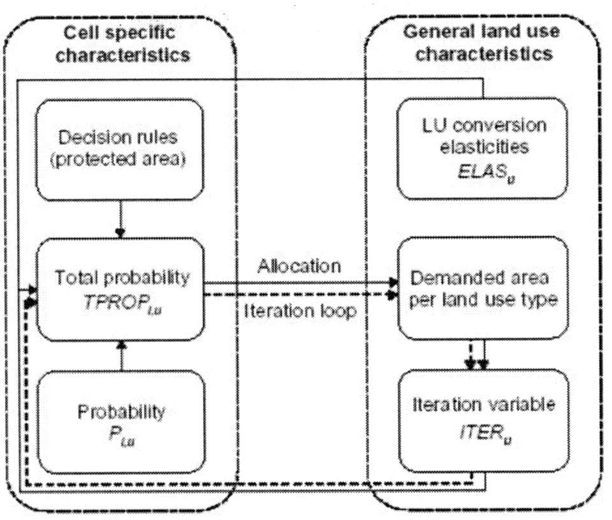

Figure 2.4. Representation of the iterative procedure for land use change allocation.

Multi-Scale Characteristics

One of the requirements for land use change models are multi-scale characteristics. The above described model structure incorporates different types of scale interactions. Within the iterative procedure there is a continuous linkage between macro-scale demands and local land use suitability. When the demand changes, the iterative procedure will ensure that the land use types for which demand increased have a higher competitive capacity (higher value for ITERu) to ensure enough allocation of this land use type. Instead of only being determined by the local conditions, captured by the logistic regressions, it is also the regional demand that affects the actually allocated changes. This allows the model to 'overrule' the local suitability, it is not always the land use type with the highest probability according to the logistic regression equation that the grid cell is allocated to.

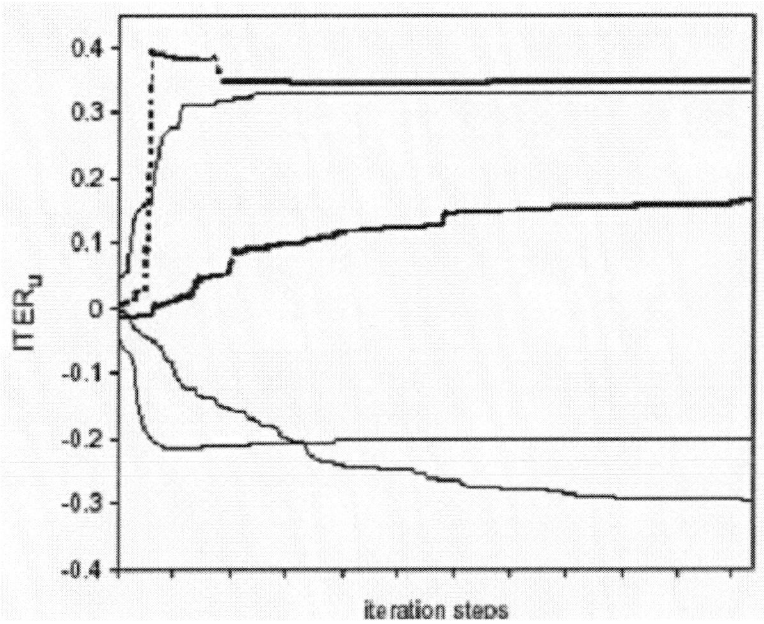

Figure 2.5. Change in the iteration paramater (ITERU) during the simulation with one time step. The different lines represent the iteration parameter for different land use types. The parameter is changed for all land use types synchronously until the alloted landuse equals the demand.

Apart from these two distinct levels of analysis there are also driving forces that operate over a certain distance instead of being locally important. Applying a neighbourhood function that is able to represent the regional influence of the data incorporates this type of variables (Figure 2.6). Population pressure is an example of such a variable: often the influence of population acts over a certain distance. Therefore the actual location of the settlements of the people is inappropriate to describe the land use pattern. The surface created by a neighbourhood function is much more capable to represent this driving factor. Instead of using these variables generated by neighbourhood analysis it is also possible to use the more advanced technique of multi-level statistics (Goldstein 1995) which enable to include higher-level variables in a straightforward manner within the regression equation (Polsky and Easterling III 2001).

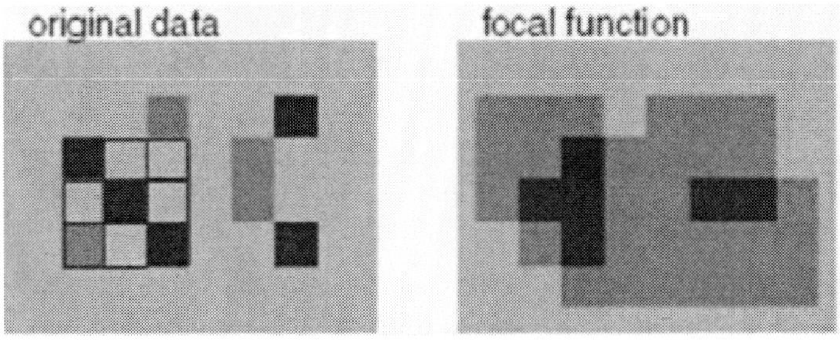

Figure 2.6. Illustration of the use of a neighbourhood function (focal function) to create a variable surface to represent the higher level influence of the variable.

2.4. Simulating Impacts on Water Resources Using SEA/BASINS and BROOKS Hydrologic Models

Water is one of the most valued resources in a watershed. It is used for a wide variety of purposes and is often referred to as one of the basic resources that drive development. However, in many watersheds water resources are now in various stages of degradation. Many rivers invariably exhibit extreme low and peak flows that make water resources management a difficult task. As commonly believed, the degradation of water resources is largely attributed to the deterioration of the watershed in general and of the land in particular brought about by improper management of land use and land use practices. Climate change and variability, specifically that of rainfall and temperature, aggravates the adverse impacts of land use. Water resource is also one of the most difficult to manage owing to the influence of factors such as climate that is beyond the direct influence of management. Further, water is also highly influenced by land use and land use practices that in many watersheds are often impossible to limit, more so to prohibit.

Impact assessement of water resources is done by using SEA/BASINS Integrated Modeling System and BROOK Hydrologic Models as discussed below.

2.4.1.Sea/Basins

SEA/BASINS Integrated Hydrological Modeling System is developed by Southeast START Research Center (SEA START RC). It uses the Variable Infiltration Capacity (VIC) Model developed by Liang et al. (1994), and the Dynamic Routing Model developed by Lohmann et al. (1996). The VIC Model (vertical) calculates the water balance at each individual grid cell while Dynamic Routing Model (horizontal) routes the runoff generated by each grid cell downstream. These two separate models have a number of advantages. It separates the 'indirect water routing' and 'direct water diversions.' The former, which include impacts of land use change and climate change, expresses them mainly through the 'vertical' model, that is, the water balance at the grid cell level. The latter, including increased withdrawals and diversions for agricultural, industrial, and domestic use, impacts mainly on the 'horizontal' model, which represents flow routing. The separation into the grid cell and channel components also allows for an easy interface to treat non-point source and in-channel chemical processes separately. The data requirements for these two models are enumerated in Table 2.2. Pre-processing and process flow for these models are described below and illustrated in Figures 2.12– 2.13.

Table 2.2. Data needed for the VIC and Dynamic Routing Models

Data for VIC Model	Data for Dyanmic Routing Model
Forcing Data Precipitation (mm) Maximum temperature (0C) Minimum temperature (0C) Windspeed (m s-1)	Elevation Stream network Dicharge data
Template Data Soil property Land cover Elevation	

In the pre-processing stage watershed boundary, stream networks, contour lines, soil and vegetation maps are processed and converted into grid maps using ArcView and are then converted to ASCII files using ArcInfo. Daily precipitation, daily wind speed, maximum and minimum temperature from three stations will be interpolated using SURFER program and GSMAC.

After pre-processing and parameterization, VIC Model will be executed to create runoff data and will be routed using Dynamic Routing Model to simulate discharge values. To make routed discharge cohere with observed discharge, simulated discharge will be calibrated by editing fraction of runoff and baseflow.

Vic Model

The VIC Model is a semi-distributed, grid-based hydrological model which parameterizes the dominant hydrometeorological processes taking place at the land surface–atmosphere interface. A mosaic representation of land surface cover, subgrid parameterizations for infiltration, and the spatial variability of precipitation account for subgrid scale heterogeneities in key hydrological processes.

The model uses two soil layers and one vegetation layer with energy and moisture fluxes exchanged between the layers (Figure 2.7). The VIC hydrology model will be coupled to a water management model to predict the effect of water withdrawal, irrigation, and reservoir operation decisions on downstream flows.

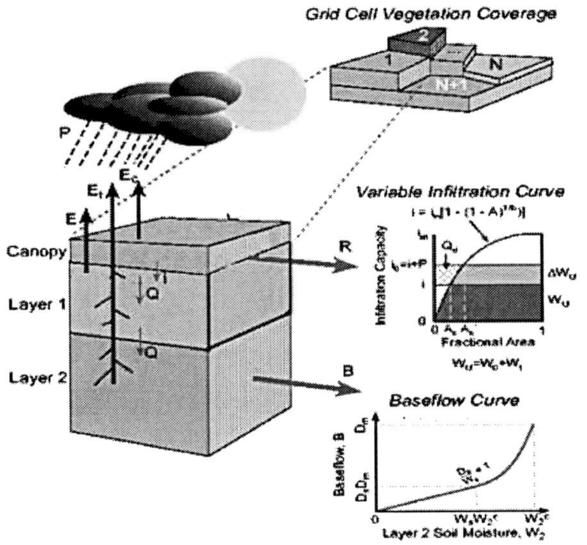

Figure 2.7. VIC Model's representation of surface and subsurface hydrologic processes based on a grid.

The VIC Model represents surface and subsurface hydrologic processes on a spatially-distributed (grid cell) basis. In typical SEA/BASINS applications, grid cells of 1 km have been used. Although each grid cell may contain different vegetation classes, for example, 33% with tall coniferous trees and 36% with grassland each represented by a different vegetation parameter set (e.g., leaf area index, surface roughness), in normal SEA/BASINS operation homogenous grids are used. Grid cell subsurface processes are represented using average soil characteristics for the entire cell. Water balance terms are computed independently for each coverage class (vegetation and bare soil) present in the model. Energy balance is excluded from normal SEA/BASINS runs.

Vegetation and soil characteristics associated with each grid cell are reflected in sets of vegetation and soil parameters. Parameters for vegetation types are specified in a user defined library of vegetation classes (usually derived from standard, national classification schemes) while the distribution over the gridded land surface area is specified in a vegetation parameter file. Soil characteristics (e.g., sand and clay percents, bulk density) are represented as two vertical soil layers (VIC-2L). The sources of data and specification for each vegetation and soil parameter are described in greater detail in the links following from the VIC Model operations web page.

Processes governing the flux and storage of water in each cell-sized system of vegetation and soil structure include evaporation from the upper soil layers, evapotranspiration, canopy interception, evaporation, infiltration, percolation, runoff and baseflow. A full discussion of the algorithms relating to these processes may be found at http://www.hydro. washington. edu/Lettenmaier/Models/VIC/VIChome.html, particularly Liang et al. (1994), Liang et al. (1998) and Cherkauer et al. (1998).

Features of interest include the eponymous variable infiltration curve, shown to the right of the soil and vegetation column schematic (Figure 2.7), which scales the maximum infiltration by a non-linear function of fractional grid cell area to enable runoff calculations for sub grid-scale areas (such as what arises from the use of multiple vegetation classes). Another feature is the specification of baseflow as a function of soil moisture in the lowest soil layer. This relationship is non-linear at high soil moisture contents, producing rapid baseflow response in wet conditions. Below a user-specified value of soil moisture, the function becomes linear, thereby reducing the responsiveness of baseflow in dry conditions. As well as for many other processes in the VIC-2L model, the parameters that define both the infiltration curve and the baseflow curves are user-specified, affording the model user a good deal of control over these processes.

The model is run one grid cell at a time over a desired period (any subset of the period spanned by the model forcing data), to produce time series of runoff, baseflow, evaporation, and other physical variables for each grid cell.

Dynamic Routing Model

The routing model was developed by Dag Lohmann. The model transports grid cell surface runoff and baseflow produced by VIC Model within each grid cell to the outlet of that grid cell then into the river system. In this model, it is assumed that water of each grid can flow to one of eight directions depending on elevation gradient (Figure 2.8). It will follow the steepest gradient downstream. The Dynamic Routing Model sums runoff from each grid to water station by taking into consideration distance and time, and exits a cell in a single flow direction.

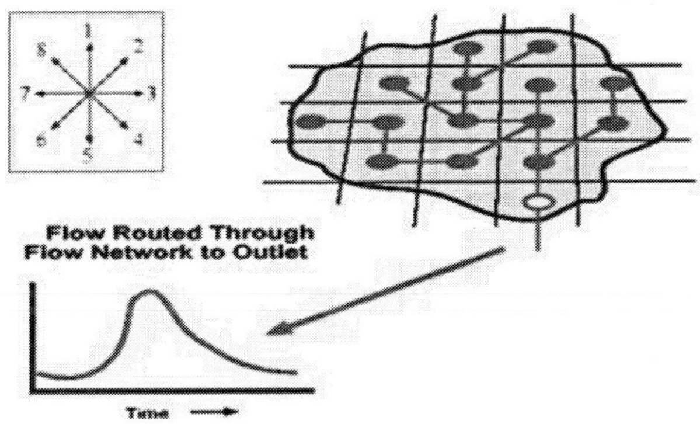

Figure 2.8. Illustration of how the Dynamic Routing Model routes runoff following the steepest gradient.

2.4.2. Pre-Processing for SEA/BASINS System

Data Preparation

Preparation for running the VIC Model consists of several pre-processing steps. These steps are the following:

1. Delineate watershed boundary using ArcView. The coverage is converted to ASCII file (mask.asc) using ArcInfo.
2. Interpolate observed daily precipitation, wind speed, maximum temperature and minimum temperature from local stations using SURFER program with two GS Scripts developed by SEA START RC (G_W.BAS and N_X.BAS). This produces

four directories of text files for the four forcing parameters (GPCP, WIND, MAX and MIN).

3. Prepare gridded integrated elevation file that contains elevation of watershed boundary (line), stream network and contour lines. ArcView is used to convert these three objects from line to grid, use eleva.aml script in ArcInfo to merge these three grid coverages, and export to eleva.asc file.
4. Develop gridded routing network, gridded flow accumulation (flowaccu) and grid identification coverages (id) from gridded integrated elevation file using routing.aml script of ArcInfo. The id coverage will be used to locate the hydrological stations for calibration/verification. The routing.aml script will also export these ascii files necessary for routing model step; i.e., velo.asc, diffu.asc, xmask.asc, flowdirec.asc, and frac.asc.
5. Prepare soil parameter file from local soil texture (%sand and %clay) data for 0–30 cm and 30–100 cm layers. Use ArcView to convert polygon into grid. Use soil.aml script in ArcInfo to convert these grid coverages into sand30.asc, sand100.asc, clay30.asc, and clay100.asc files.
6. Convert vegetation (land cover) classification into UMD classification (13 classes) system and converting vegetation polygon to grid coverage using ArcView. Using veg.aml of ArcInfo to convert the gridded vegetation into veg_mask.asc.

Parameterization for SEA/BASINS Execution

7. Set up location of forcing related files by specifying the locations of forcing data files (Step 2) and create daily forcing for each grid cell. Specify the beginning and end dates for the run.
8. Create soilfile by specifying the name and location of soil data file (Step 5). Soil parameter can also be adjusted manually in the soil parameter calibration menu so that the calculated soil moisture agrees with the observed data at the same location. The program will also automatically call elevation.asc (Step 3), and mask.asc (Step 1) files. The file id.asc will also be created in this step.
9. Create vegfile from veg_mask.asc (Step 6), id.asc (Step 4) and vegetation index (veg_inde.asc) which is supplied with the program (for UMD classification).
10. Specify the starting day, month, year and ending day, month, year that VIC-2L will be run.

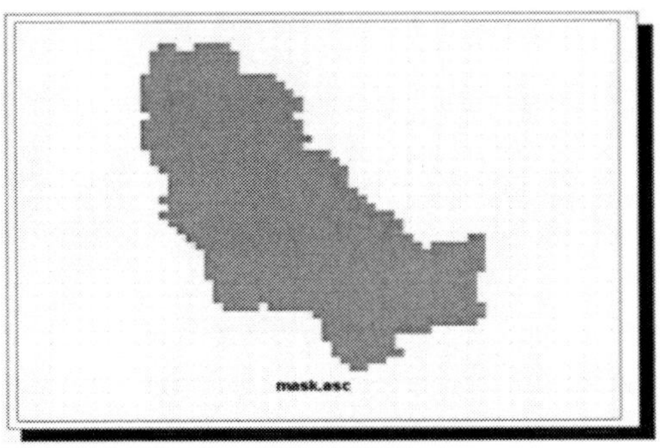

Figure 2.9. Snapshot of masc.asc file.

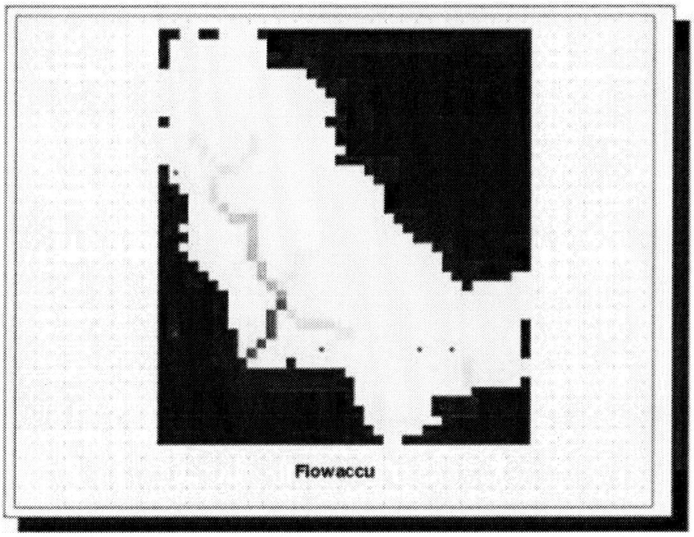

Figure 2.10. Snapshot of flowaccu file.

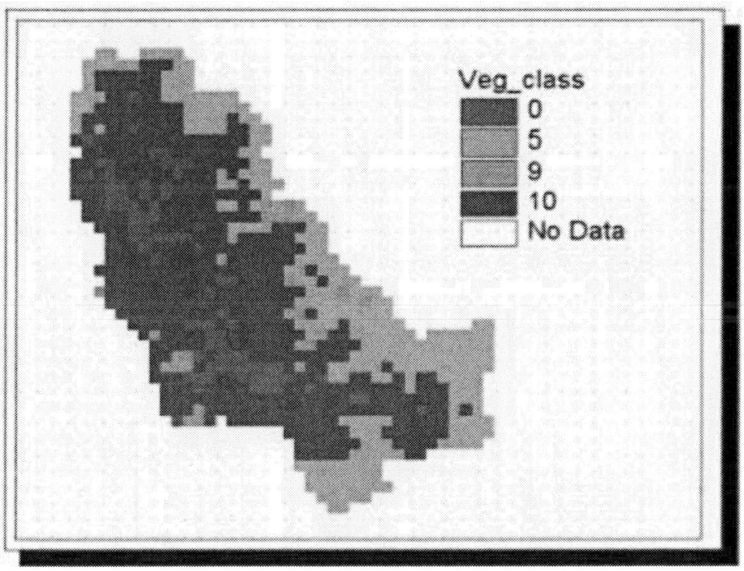

Figure 2.11. Snapshot of vegetative cover in grid format.

2.4.3. Execution of VIC Model

1. Run VIC-2L using forcing (Step 7), soilfile (Step 8), vegfile (Step 9) and vegetation library (veg_lib) which is supplied with the program (for UMD classification).
2. Further refining VIC run, if desired, can be done by editing soilfile for these variables: variable infiltration curve parameter (infilt), or fraction of maximum velocity of baseflow where non-linear baseflow begins (Ds), maximum velocity of baseflow (Dsmax), or fraction of maximum soil moisture where non-linear baseflow occurs (Ws), or average soil temperature (avg_T), or soil thermal damping depth

(dp), or fractional soil moisture content at the critical point (Wcr_FRACT), or fractional soil moisture at the wilting point (Wpwp_FRACT), surface roughness of baresoil (rough), or surface roughness of snow pact (snow_rough), or average annual precipitation (annual_prec).

2.4.4. Execution of Flow Routing Model

1. Locate the identification number (ID) of the grid cell that flow will be routed to from its latitude and longitude using ArcView and gridded flow accumulation coverage (Step 4). Put the name of the station and its gridded ID in the appropriate boxes in the Routing Model window.
2. For model calibration, enter the time period when that observed daily discharge data is available at that station in appropriate boxes. Calibration can be done by adjusting the fractions of runoff and of baseflow to be routed into channels until the routed discharge agrees with observed discharge at that observation station.

Once the optimum fractions of runoff and baseflow to be routed are obtained, use these fraction to calculate discharge for other time periods as long as forcing data are available.

2.4.5. Brook Hydrologic Models

The BROOK Hydrologic Models simulate the water budget on a unit land area at a daily time step. It is applicable to the land phase of the precipitation-evaporation-streamflow part of the hydrologic cycle for a point or for a small, uniform (lumped parameter) watershed. The BROOK 90 uses Penman-Monteith equation to compute for the potential evapotranspiration while BROOK 5 uses the Hamon method (Federer 1995). Tables 2.3 and 2.4 enumerate the data needed to run the BROOK 90 and BROOK 5 models, respectively.

Table 2.3. Data needed to run the BROOK 90 Model

Parameter	Unit
Area	ha
Latitude	
Daily rainfall (years of record)	mm
Daily maximum and minimum temperature	°C
Daily solar radiation	MJ m-2
Daily average vapor pressure	Kpa
Daily average wind speed	m s-1
Streamflow	

Table 2.4. Data needed to run the BROOK 5 Model

Parameter	Unit
Area	ha
Latitude	
Highest elevation	m
Lowest elevation	m
Basin length	m
Direction of streamflow	
Land use	
% of Forest	
% of Crops and pasture	
% of Bldg/roads	
% of Bodies of water	
Water table depth	m
Daily mean temperature (years of record)	
Daily rainfall and (years of record)	
Soil texture	
Drift deposits	

2.5. Multi-Level Vulnerbality Indicator of Local Communities In PCW

Local communities living within watersheds are subject to major natural occurrences that reflect climate variability and climate change over the last few decades. Usually, communities are composed of different socioeconomic groups with varying degrees of vulnerability. Within the watersheds, too, can be found different user-institutions which may exert socioeconomic influence over the vulnerability of local communities in relation to climate variability and extremes. This part of the study discusses methods drawn from the Philippine experience that could help advance the current state of knowledge and policies relevant to the vulnerability of local communities and institutions to climate variability and extremes.

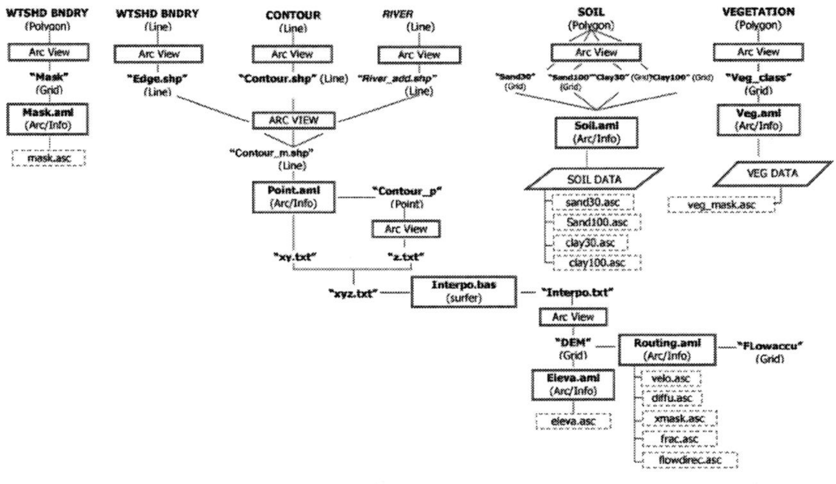

Figure 2.12. Pre-processing of data for SEA/BASINS Modeling System.

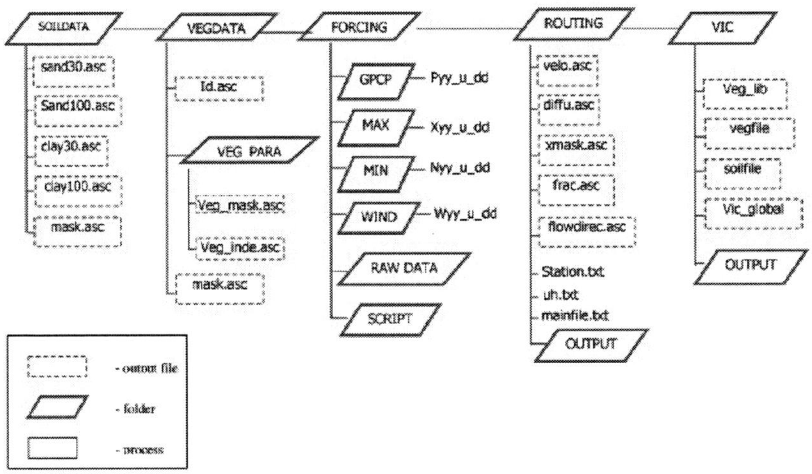

Figure 2.13. Process flow of SEA/BASINS Modeling System (VIC Model and Dynamic Routing Model).

2.5.1. Data Collection

The study employs a combination of data collection methods: secondary data gathering, household survey, use of participatory rural appraisal techniques, direct field observation, and GPS readings of identified vulnerable areas. The following sections briefly discuss these methods.

Secondary Data Gathering

Available secondary information on the biophysical and socioeconomic aspects of the watershed are gathered from relevant agencies to understand the local and regional context of the watershed communities. Sources of information include municipal and provincial development plans, sociodemographic statistics from the National Statistics Office, atlas and other maps from various sources, project documents, and other pertinent information from different institutional stakeholders of PCW.

Meanwhile, climatic data like rainfall, temperature, El Niño and La Niña episodes, and other natural calamities that occurred in PCW are obtained from PAGASA and the weather station near the watershed. These data are gathered on a historical basis.

Household Survey

Household survey is conducted to determine the vulnerability of households to climate variability and extremes and the socioeconomic factors influencing their vulnerability. It makes use of a pre-tested interview schedule that contains the following information: (1) socioeconomic profile of the respondent; (2) household's use and benefits from PCW; (3) climate variability and extremes experienced in the last few decades and their impacts; (4) household's vulnerability in terms of food availability, water supply, livelihood, and health; and (5) adaptation strategies.

The survey covers the four municipalities of the three different provinces encompassing the watershed. These are Pantabangan and Carranglan in Nueva Ecija, Alfonso-Castañeda in Nueva Vizcaya, and Ma. Aurora in Aurora. Twenty-six of the 36 barangays within the watershed area are covered. Ten of 36 barangays are excluded since a very small portion of their respective areas is within the watershed boundary, hence, very few people live in these areas. A total of 375 respondents are randomly selected using the barangay records. This sampling technique employed is adopted from Chua (1999) which allows a 0.05 permissible error and 95% confidence interval level.

PRA Techniques

Focus groups discussions (FGDs) are done in 21 barangays to complement the household survey and determine the vulnerability of various socioeconomic groups at the community level. The FGDs employed a combination of participatory rural appraisal (PRA) techniques such as time line analysis, stakeholder analysis, participatory vulnerability assessment, and community mapping. The choice of these techniques was guided by the different research questions, the explicit objective of the study to engage the local stakeholders in the process of assessing their current vulnerability, and the literacy level of the local communities (see Pulhin 2002 for discussions on these techniques).

Field Observations and GPS Readings

Direct field observation is also conducted to validate information gathered through household survey and focus group discussions. In addition, GPS readings of vulnerable areas identified by the local communities are also collected for purposes of mapping these areas.

2.5.2. Data Analysis

The study employs a combination of qualitative and quantitative approaches to analyze the information gathered through the abovementioned methods. Qualitatively, the degree of present vulnerability of the different socioeconomic groups is assessed by aggregating and analyzing the results of FGDs. At the household level, a more quantitative technique using correlation and regression analyses are employed to determine the factors influencing the household's vulnerability based on the vulnerability index developed. In addition, vulnerable areas are identified using GIS to complement the participatory vulnerability mapping conducted by the local communities.

Qualitative Analysis

Results from FGDs conducted in 21 barangays are combined and synthesized to identify the climate variability and extremes experienced by the local communities in the last few decades, to determine the more vulnerable groups and their location in the watershed, and to assess the nature and extent and of their vulnerability. This qualitative analysis centers on the vulnerability situation of major socioeconomic groups in the watershed as identified by the community members themselves during the FGDs. The emphasis on socioeconomic groups provides a broader perspective of community vulnerability and complements the more micro and quantitative analysis done at the household level.

Development of Vulnerability Index

Results from the household survey are used to develop a vulnerability index. The index consisted of four major component indicators: food, water, livelihood, and health. The indicators are further divided into subcategories, each of which are given corresponding weights. Drawing from the framework of Moss et al. (1999), the subcategories comprises relevant variables that involve certain characteristics of the component indicators in relation to climate variability and extremes (representing the household's sensitivity in relation to these components) and the presence or absence of adaptation strategies (representing the household's coping capacity).

Two types of weights are considered in the development of the index: researchers' weights representing the experts' view, and that of the watershed communities' representing the local stakeholders' view. The first iteration for the vulnerability index computation is based on the researchers' judgment that makes use of composite weighing where all the four major components (food, water, livelihood, and health) are given equal weights (25 points each) with a grand total of 100 points. The subcategories under each major component are also given corresponding weights with each level of the subcategories given equal points.

Other than the researchers' judgement, the local communities' perspective is also taken into account in the development of the index. Using the same set of indicators developed by the researchers, two separate FGDs are conducted in two clusters of barangays in the municipalities of Pantabagan and Carranglan where participants are asked to provide their own weights for the index. Consensus is sought from the participants during the FGDs on specific weights that they should assign for each component indicators at various levels.

Table 2.5 presents the vulnerability index developed using the researchers' judgment and that of the local communities'.

Correlation and Regression Analysis

The computed final vulnerability index, food index, water index, livelihood index and health index are correlated with the factors hypothesized to influence vulnerability using Spearman Correlation.

Table 2.5. Multi-level indicator of vulnerability of PCW households to climate variability and extremes (CV and E) using varying weights

Vulnerability index	Weights by researchers	Weights by local communities		
		P	C	P and C
A. Food	25	25	40	32.5
a. 1 Seed availability	12.5	20	15	17.5
a. 1.1 Availability of planting materials	4.17	8	7	7.5
i. Available anytime of the year	0	3	2	2.5
ii. Seasonal or hard to find	4.17	5	5	5
a.1.2 Is it affecter by CV and E?	4.17	9	5	7
i. Yes	4.17	9	4	6.5
ii. No	0	0	1	0.5
a.1.3. Adaptation strategies	4.17	3	3	3
i. With adaptation	0	2	1	1.5
ii. Without adaptation	4.17	1	2	1.5
a.2. Crop Yield	12.5	5	25	15
a.2.1. Percent lost in rice production	4.17	1.5	10	5.75
a.2.2 Is it affected by CV and E?	4.17	2	10	6
i. Yes	4.17	2	7	4.5
ii. No	0	0	3	1.5
a.2.3 Adaptation strategies	4.17	1.5	5	3.25
i. With adaptation	0	0.5	2	1.25
ii. Without adaptation	4.17	1	3	2
		P	C	P and C
B. Water	25	40	40	40
b.1. Domestic water	12.5	33	15	24
b.1.1 Sources of domestic water	2.5	11	7	9
i. Natural sources	2.5	8	6	7

Table (Continued)

ii. Through agencies	1.25	3	1	2
b.2.1 Distance of house to sources of water	2.5	5	2	3.5
i. 0– 250 m	0.62	0.4	0.2	0.3
ii. 251 – 500 m	1.25	1	0.3	0.65
iii. 501 – 1000 m	1.88	1.5	0.5	1
iv. > 1000 m	2.5	2.1	1	1.55
b.1.3 Observation for the supply of domestic water	2.5	7	2	4.5
i. Declining supply	2.5	3	1	2
ii. Increasing supply	0	2	0.5	1.25
iii. No change	1.25	1	0.5	0.75
b.1.4 Is domestic water supply affected by CV and E?	2.5	5	2	3.5
i. Yes	2.5	3	1.5	2.25
ii. No	0	2	0.5	1.25
b.1.5 Adaptation strategies	2.5	5	2	3.5
i. With adaptation	0	1	0.5	1.25
ii. Without adaptation	2.5	4	1.5	2.75
b.2 Irrigation water	12.5	7	25	16
b.2.1 Regularity / problem with supply?	4.17	3	10	6.5
i. Problem with supply	0	1	3	2
ii. No problem with supply	4.17	2	7	4.5
b.2.2 Effects of scarcity	4.17	2	10	6
i. Decrease in production /income	2.78	1	7	4
ii. No (zero) production / income	4.17	0.5	1	0.75
iii. Delayed harvest	1.39	0.5	2	1.25
b.2.3 Adaptation strategies	4.17	2	5	3.5
i. With adaptation	0	0.56	2	1.28
ii. Without adaptation	4.17	1.44	3	2.22
C. Livelihood	25	15	10	12.5
c.1 Seek sources of income in cases of CVandE?	8.33	6	2	4
i. Yes	0	4	0.5	2.25
ii. No	8.33	2	1.5	1.75
c.2 Is income from other sources sufficient?	8.33	6	6	6
i. Sufficient	0	2	2	2

ii. Not sufficient	8.33	4	4	4
c.3 Adaptation strategies	8.33	3	2	2.5
i. With adaptation	0	2	0.5	1.25
ii. Without adaptation	8.33	1	1.5	1.25
D. Health	25	20	10	15
d.1 Experienced health problems during CVandE?	6.25	6	2	4
i. Yes, experience health problems	6.25	4	1.5	2.75
ii. No	0	2	0.5	1.25
d.2 Kinds of health problems experienced during CVandE	6.25	7	4	5.5
i. Diarrhea, amoebiasis, dehydration, dysentery"	4.17	3	2	2.5
ii. Dengue, typhoid, malaria	6.25	2	1	1.5

Legend: (P) Pantabangan; (c) Carranglan; (P and C) Pantabangan and Carranglan.

These include a combination of demographic, socioeconomic, and geographic factors including the number of coping mechanisms practiced by each household. Moreover, to determine the combined effects of the different hypothesized factors on households' vulnerability, regression analysis is done using SPSS ver. 10 for Windows. Both the correlation and regression analysis used a 0.01 to 0.05 level of significance.

Mapping of Vulnerable Areas

The vulnerability of PCW to climate variability and extremes is assessed using an arbitrary set of rules (Table 2.6) related to five key parameters, namely: slope, elevation, distance from the road, distance from the river and distance from the community center. With the aid of GIS, the degree of vulnerability by land use type is determined for the entire watershed using the category of low, medium, and high vulnerability. A single vulnerability map is developed by overlaying all the individual maps produced for each of the five parameters.

On the other hand, GPS readings are made for all the vulnerable places identified by the local communities themselves during the FGDs conducted in the different barangays using the participatory vulnerability mapping technique. The GPS readings of the vulnerable places are plotted in the vulnerability map of the watershed developed through GIS. The idea is to determine whether there will be congruence between vulnerable areas identified using biophysical parameters through GIS with what the stakeholders see as vulnerable places.

Table 2.6. Criteria used for the assessment of vulnerability of PCW to climate change

Landuse	Slope			Elevation (m)			Distance from road (m)			Distance from river (m)			Distance from community (km)		
	Low	Mod	High	Low	Mod	High	Low	Mod	High	Low	Mod	High	Low	Mod	High
Grass /Brush	0-50	-	>50	100-250	250-500	>500	>500	200-500	<200	>1000	500-1000	<500	>1000	500-1000	<500
Agriculture	<8	8-18	>18	100-250	250-500	>500	<500	500-1000	>1000	<500	500-1000	>1000	<500	500-1000	>1000
Forests	<18	18-50	>50	100-250	250-500	>500	>1000	500-1000	<500	>1000	500-1000	<500	>1000	500-1000	<500

Table 2.7. Data requirements for the socioeconomic component of the study

Data Category	Time Scale					Time Resolution			Spatial resolution/level			
	Historical	Present	2020	2050	2100	Multi-decadal	Annual	Seasonal	Farm/ household	Community	Watershed	National
Demographic trends												
Population trend	X	X	X	X	X	X				X	X	X
No. and size of households	X	X	X	X	X	X				X	X	X
Population density	X	X	X	X	X	X				X	X	X
Annual growth rate	X	X	X	X	X	X	X			X	X	X
Migration pattern	X	X	X	X	X	X				X	X	X
Age and sex composition	X	X	X	X	X	X				X	X	X
Fertility and mortality rate	X	X	X	X	X	X				X	X	X
Ethnic groups	X	X	X	X	X	X				X	X	X
Socioeconomic trends/characteristics												

Sources of livelihood		X	X	X	X		X	X	X	X	X	X
Income and expenditures		X	X	X	X		X	X	X	X	X	X
Educational attainment		X	X	X	X		X	X	X	X	X	X
Data Category	Time Scale					Time Resolution			Spatial resolution/level			
	Histo-rical	Present	2020	2050	2100	Multi-decadal	Annual	Seasonal	Farm/ household	Community	Watershed	National
Employment/ unemployment	X	X	X	X	X		X	X	X	X	X	X
Literacy rate	X	X	X	X	X		X	X	X	X	X	X
Labor supply	X	X	X	X	X		X	X	X	X	X	X
Land ownership	X	X	X	X	X		X	X	X	X	X	X
Food consumption	X	X	X	X	X		X	X	X	X	X	X
Demand for water	X	X	X	X	X		X	X	X	X	X	X
Demand for fuelwood	X	X	X	X	X		X	X	X	X	X	X
Demand for timber and other forest products	X	X	X	X	X		X	X	X	X	X	X
Demand for settlement and agricultural land	X	X	X	X	X		X	X	X	X	X	X
Agricultural production	X	X	X	X	X		X	X	X	X	X	X
Cultural practices/beliefs related to forests	X	X	X	X	X	X	X			X	X	X
Socioeconomic groupings	X	X	X	X	X	X				X	X	
Geographic factors												
Distance from farm	X	X	X	X	X	X				X	X	
Distance from road	X	X	X	X	X	X				X	X	
Distance from market	X	X	X	X	X	X				X	X	
Governance structure	X											

Table 2.7 (Continued)

Forest policy	X	X				X				X	X	X
Institutional arrangements		X				X				X	X	X
Government support and services		X				X				X	X	X

Table 2.8. Research questions and research techniques used in the socioeconomic component of the study

Research questions	Research method/PRA technique	Remarks
What are the natural occurrences in the area that reflect climate variability and extremes over the last 50 years or so?	Participatory Time Line Analysis of climate-related events Household interviews	Focus Group Discussions of different socioeconomic groups are conducted in the study barangays
What are the impacts of these natural occurrences on the socioeconomic well-being of the local communities?	Participatory Impact-Analysis of climate related events Household interviews	Impacts of climate variability and extremes to local communities may include impacts on water supply, crops production, livelihood, lost of lives, and properties, etc.
What are the different socioeconomic groups and institutions/agencies in the area that are affected by climate variability and extremes?	Stakeholder analysis of socioeconomic groups dependent on the watershed. Institutional analysis of different agencies concerned on watershed management	The different socioeconomic groupings of local communities are identified as well as the different government, private, and non-government institutions working in the area
Who are the vulnerable sectors in the community?	Stakeholder Analysis/Focus Group Discussions Identification of vulnerability index Household interviews	Identification of vulnerable socioeconomic groups probably based on livelihood, e.g., crop farmers tilling their own land, tenants, people whose livelihood are at risk, etc. Vulnerability index is also be developed based on literature and in consultation with the community members. Such index could include among others, household size, total income, total expenditure, crop sales price in bad year, crop land, road access, livestock holdings, social capital, household assets, etc.
What makes the different socioeconomic groups vulnerable to climate change/variability and extremes?	Household interviews, Focus Group Discussions	Identification of causal factors that influence present vulnerability

In which part of the watershed are the vulnerable groups/areas located?	Community mapping of vulnerable areas	Vulnerable areas in selected barangays are identified using participatory mapping technique with the aid of existing topographic and administrative maps with delineated barangay boundary. Vulnerable areas may be located in head waters, main tributary, downstream, etc.
What are the communities' current coping mechanisms/strategies and capacities to climate variability and extremes?	Focus Group Discussions Household interviews	Various forms of coping mechanism may include out-migration, selling of productive assets, resorting to adaptive coping strategies (e.g., curtailing consumption of some items, etc.), normal income generating pattern, etc.
What lessons can be gleaned from current vulnerability and coping capacities for adapting to future climate change?	Focus Group Discussions Household interviews	Recommendations to promote appropriate adaptation strategies to future climate change are generated from the local communities.

Chapter 3

CLIMATE CHANGE IN PANTABANGAN-CARRANGLAN WATERSHED

The first part of this chapter presents the current climatic profile of PCW while the second part presents projected climate change using various GCM scenarios.

This stream in the plains of PCW has been witness to how the watershed's climate profile has changed through the passing of decades. What is now a seemingly quiet stream may in the rainy season transform into a raging river that feeds the vast Pantabangan Dam reservoir.

One of the difficult challenges in climate change impacts assessment is downscaling GCM projections in a small country like the Philippines. GCM projections which are on a scale of hundreds of kilometers are too coarse for local impacts assessments.

For example, models used to simulate the ecological effects of climate change usually operate at spatial resolutions varying from a single plant to a few hectares (Mearns et al. 2003).

In this project the assistance of the Tyndall Centre (UK) helped to generate climate scenarios for PCW. These scenarios were subsequently used to assess the impacts of climate change to water resources in the watershed.

3.1. RECENT CLIMATE TRENDS

Trends in Observed Rainfall

Based on records from two synoptic stations of PAGASA, the annual rainfall in PCW ranges between less than 1200 mm to more than 2500 (Figure 3.1). In general, the annual rainfall appears to be stable as far as records from 1960 to 2001 are concerned. Over the last decade however, a slightly decreasing trend in the annual rainfall is noted from the Cabanatuan station and a slightly increasing trend is noted in the Central Luzon State University (CLSU) station (Figure 3.2 and Figure 3.3).

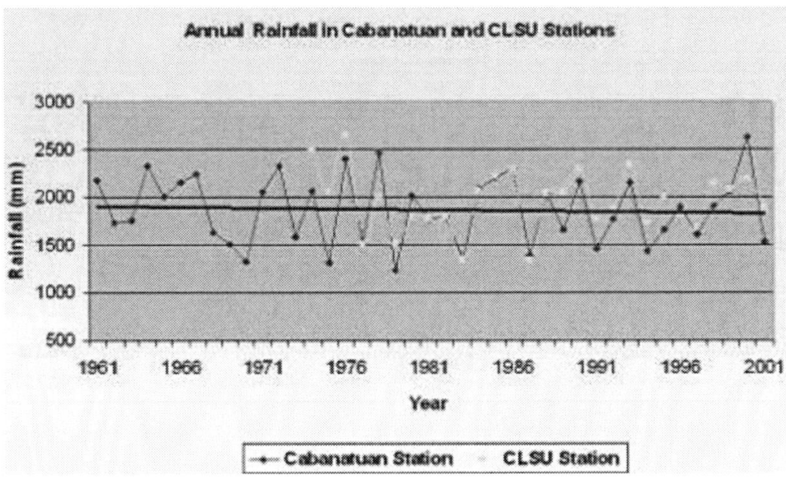

Figure 3.1. Comparison of annual rainfall in Cabanatuan and CLSU stations.

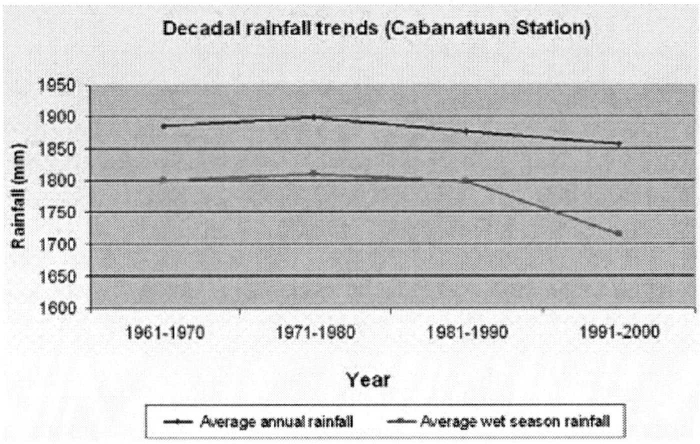

Figure 3.2. Decadal rainfall trends in Cabanatuan station.

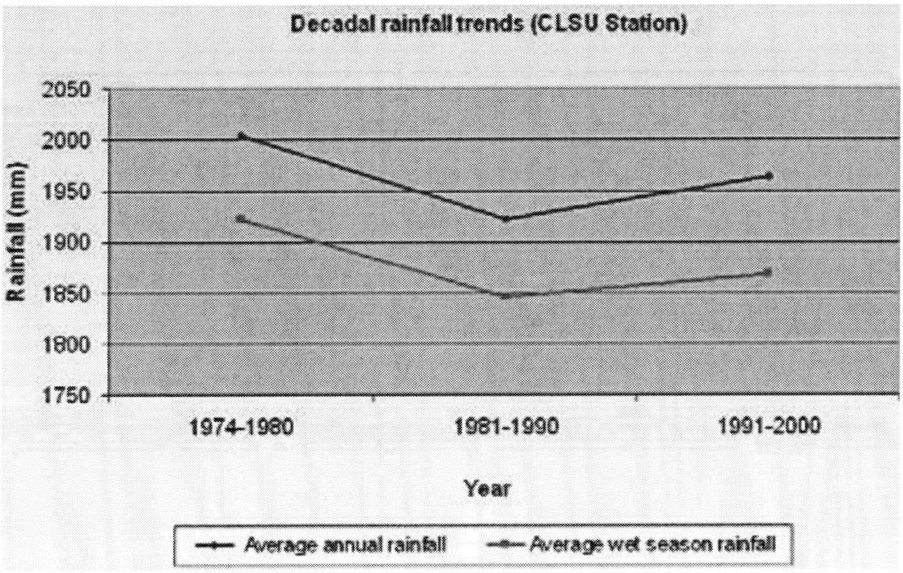

Figure 3.3. Decadal rainfall trends in CLSU station.

The number of rainy days seems to be increasing especially starting in the late 90s onto the new millennium (Figure 3.4) even as no significant perceptible change in the trend of the number of consecutive rainy days was noted (Figure 3.5). As for the maximum rainfall events a decreasing trend after 1981 is noticeable, although it is uncertain if such trend is due to climate change or change in the mode and circumstance of monitoring rainfall (Figure 3.6).

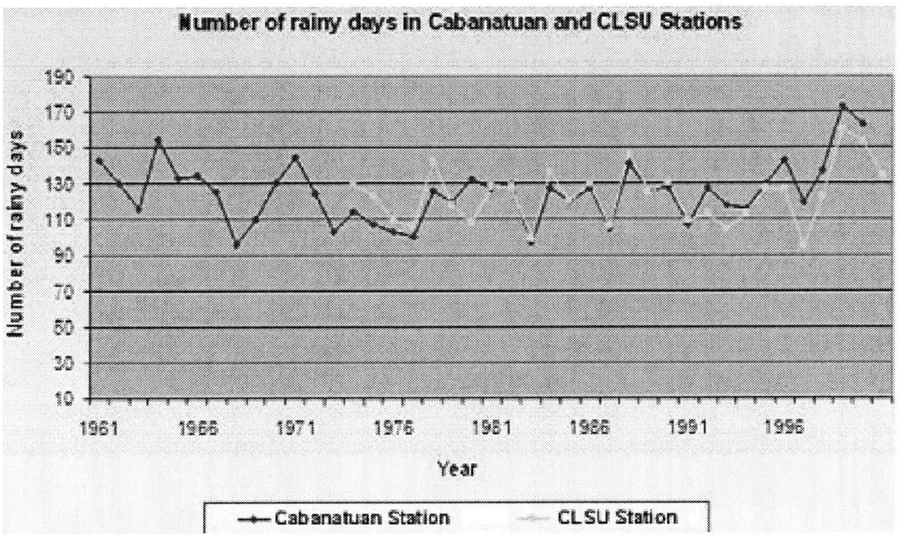

Figure 3.4. Number of rainy days in Cabanatuan and CLSU stations.

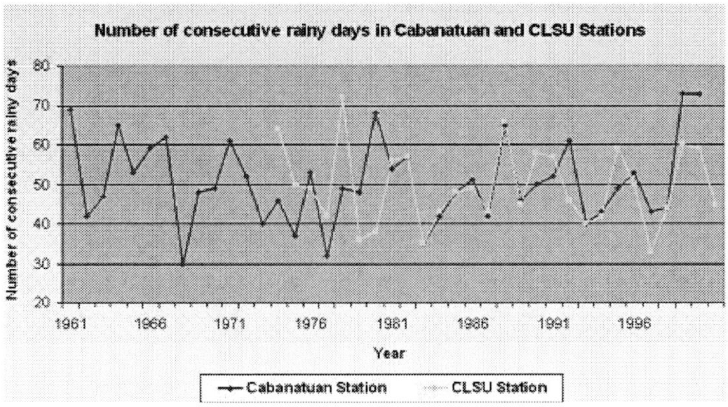

Figure 3.5. Number of consecutive rainy days in Cabanatuan and CLSU stations.

The onset of rainy season usually falls between May and June which is normal for the kind of climate in PCW. There is, however, a slight tendency for the onset of the rainy season to come earlier than May toward the end of the century (Figure 3.7). Again, it is yet too early to pronounce if this constitutes a change in the pattern of arrival of the rainy season that will persist over the long term or merely a short term variability pattern.

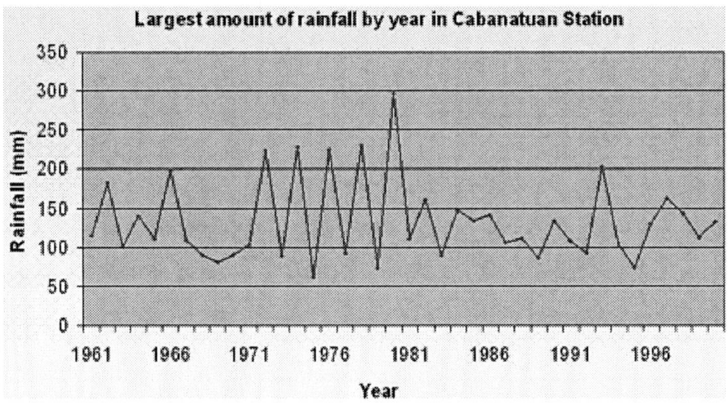

Figure 3.6. Largest amount of rainfall by year in Cabanatuan station.

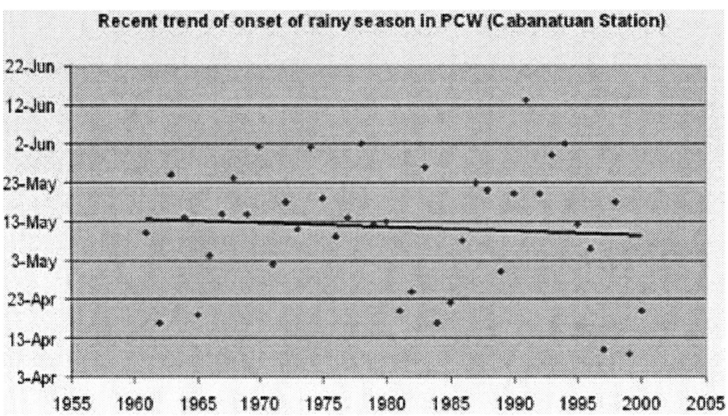

Figure 3.7. Recent trend of onset of rainy season in PCW (Cabanatuan station).

Trends in Observed Temperature

All parameters of temperature, i.e., annual mean temperature, maximum and minimum temperature, and number of days temperature is above normal were observed to demonstrate a tendency to increase over time (Figure 3.8–3.13). On the average the maximum temperature increased by 0.22 0C over the last 30 years. Increases in monthly maximum range between a low of 0.2 0C in January to as high as 2.4 in February for CLSU station. For Cabanatuan station the annual maximum temperature increase is 0.24 0C for the last 40 years, with monthly values increasing from as low as 0.27 0C in October to as high as 0.89 0C in January. For minimum temperature, a yearly average increase of 0.06 0C was noted with the monthly minimum temperature increasing between 0.12 0C in June and 0.77 0C in December.

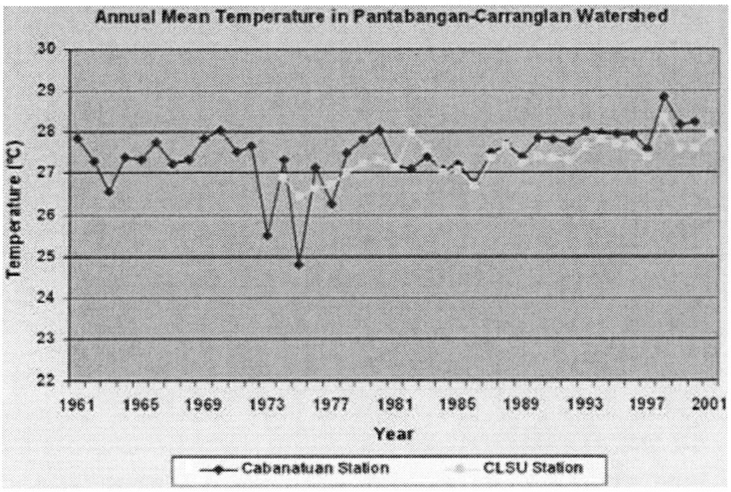

Figure 3.8. Annual mean temperature in PCW.

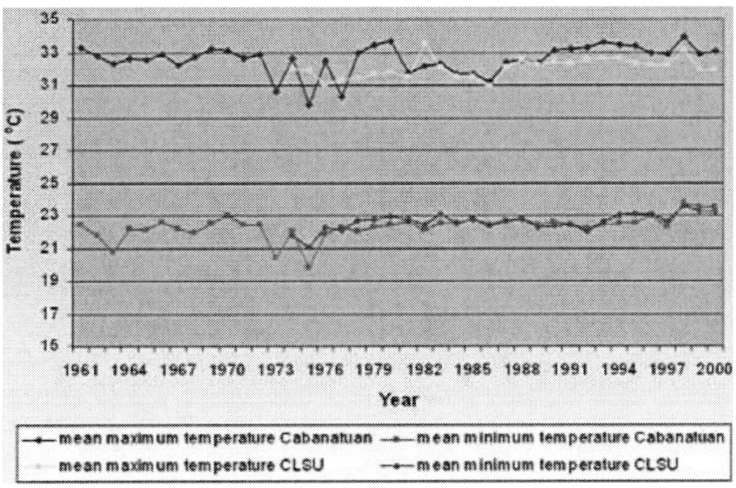

Figure 3.9. Comparison of annual mean minimum and mean maximum temperature in PCW.

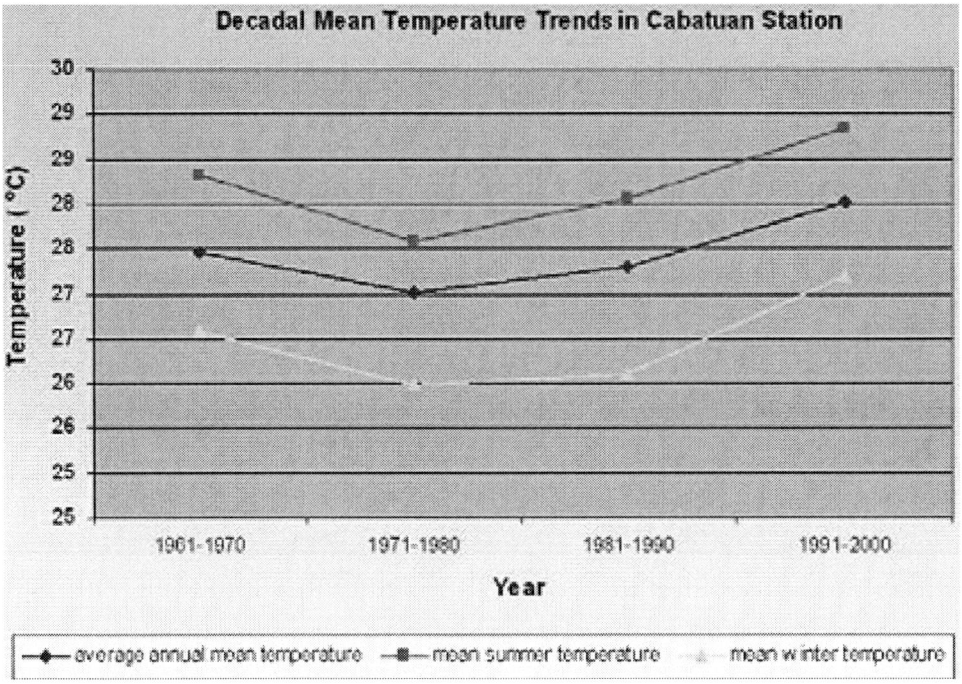

Figure 3.10. Decadal mean temperature trends in Cabanatuan station.

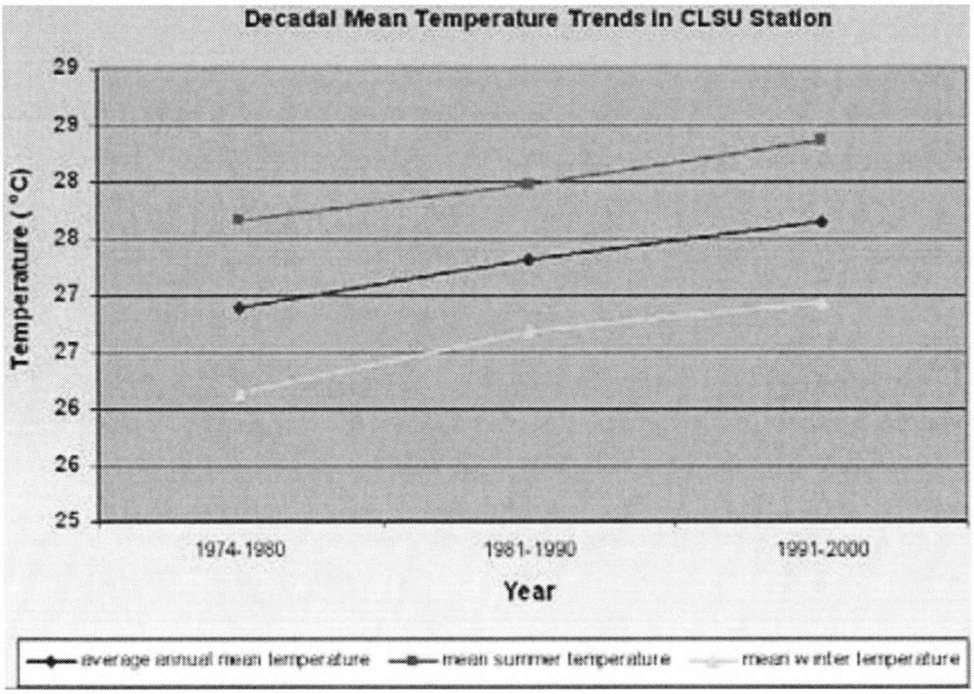

Figure 3.11. Decadal mean temperature trends in CLSU station.

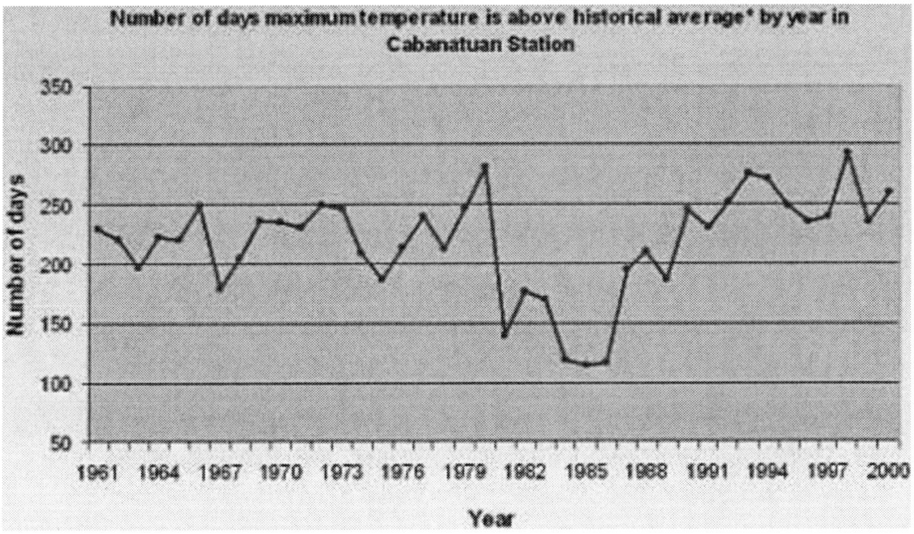

Figure 3.12. Number of days maximum temperature is above historical* average in Cabanatuan station. (*Historical average is the average of annual rainfall from 1961-2000).

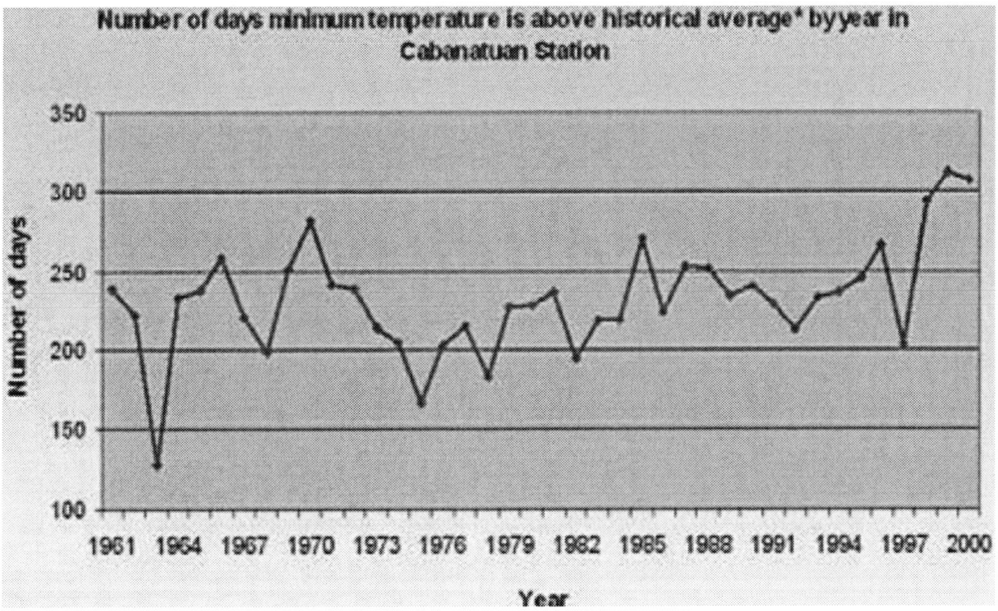

Figure 3.13. Number of days minimum temperature is above historical* average in Cabanatuan station. (*Historical average is the average of annual rainfall from 1961-2000).

Recent Trends in ENSO Events

Over the last 20 years (1980-2000) increasing frequency in the occurrence of El Niño-Southern Oscillation (ENSO) events is noted (3.1). Two of the strongest ENSO events are recorded during the same period (1982-83 and 1997-98).

Table 3.1. Occurrence of El Niño and La Niña episodes through the years

Year	1st Q	2nd Q	3rd Q	4th Q
1950	C	C	C	C
1951	C			W-
1952				
1953		W-	W-	
1954			C-	C
1955	C	C-	C-	C+
1956	C	C	C	C-
1957		W-	W-	W
1958	W+	W	W-	W-
1959	W-			
1960				
1961				
1962				
1963			W-	W
1964			C-	C
1965	C-		W	W+
1966	W	W-	W-	
1967				
1968				W-
1969	W	W-	W-	W-
1970	W-			
1971	C	C-	C-	C-
1972		W-	W	W+
1973	W		C-	C+
1974	C+	C	C-	C-
1975	C-	C-	C	C+
1976	C			W-
1977				W-
1978	W-			

1979				
1980	W-			
1981				
1982		W-	C	W+
1983	W+	W		C-
1984	C-	C-		C-
1985	C-	C-		
1986			W-	W
1987	W	W	W+	W
1988	W-		C-	C+
Year	1st Q	2nd Q	3rd Q	4th Q
1989	C+	C-		
1990			W-	W-
1991	W-	W-	W	W
1992	W+	W+	W-	W-
1993	W-	W	W	W-
1994			W	W
1995	W			C-
1996	C-			
1997		W	W+	W+
1998	W+	W	C-	C
1999	C	C-	C-	C+
2000	C+			

Legend: (C -) weak La Niña (C) moderate La Niña (C+)strong La Niña
(W -) weak El Niño (W) moderate El Niño (W+) strong El Niño.

Trends in Observed Typhoons and other Related Calamities

Limited records of climate-related calamities show that there is an apparent decline in the number of damaging typhoons defined in terms of the amount of damages caused (Table 3.2). It is also observed that the number of damaging typhoons is less in times when there are strong ENSO events (i.e., 1982-83 and 1997-98) consistent with many observations in other regions. Drought records also bear the relation with ENSO events except in 1997-98 events when no drought was recorded on or after the event. This is likely due to the mitigating effect of the transbasin transfer of water which started in 1998.

Table 3.2. Natural calamities in PCW from 1980-2003

Year	Number of damaging typhoons	Number of drought episodes	Number of earthquakes
1970	2		
1971	1		
1972	1		
1973	2		
1974	6		
1975	0		
1976	3		
1977	4		
1978	4		
1979	4		
1980	2		
1981	2		
1982	4		
1983	1	1	
1984	2		
1985	3		
1986	3		
1987	3	1	
1988	3		
1989	7		
1990	6	1	1
1991	1		
1992	7		
1993	6		
1994	5		
1995	7		
1996	2		
1997	3		
1998	4	1	
1999	6		

2000	4		
2001	2		
2002	3		
2003	3		

Source: Unpublished reports of the provincial and regional disaster coordinating councils of Pampanga and Region III.

Trends in Observed Streamflow

Information on streamflow characteristics is essential in understanding and coping with water related disasters like floods and droughts. The rise and fall of streamflow have far ranging influences on the occurrence and magnitude of floods and droughts. Characterizing streamflow patterns is, however, a complicated process owing to the many interacting factors and processes that determine the rise and fall of streamflow. Among others, rainfall, land use and land cover, soil and topography belong to the key set of factors that influence streamflow. In this part of the study focus will be on the influence of rainfall.

Between 1980 and 2000, observed streamflow data show that as rainfall rise and fall so does streamflow. Figure 3.14 shows that the pattern of rainfall from 1980 to 2000 is very similar to the pattern of monthly streamflow averages for both the wet and dry season. This rainfall-streamflow relation has been documented in many studies such as those reported in IPCC TAR (2001). It can also be seen that the monthly observed streamflow averages are following a perceptible upward path except for the dry season flow that seems to be more static. The ENSO events appear to also affect the pattern of streamflow as the significant rise and drop of the hydrograph coincide with the strong ENSO events of 1982-83 and 1997-98.

3.2. Future Climate Trend

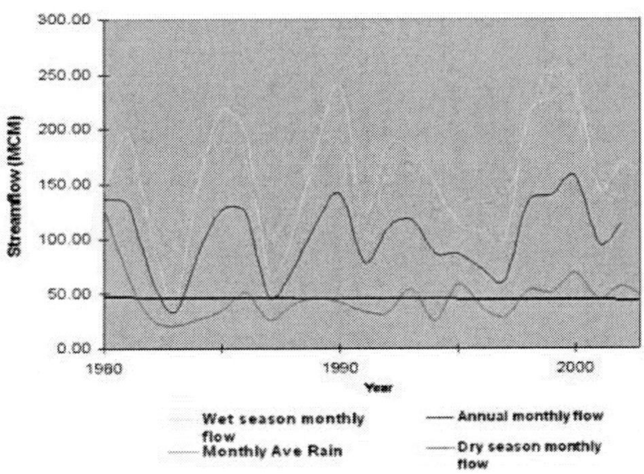

Figure 3.14. Observed monthly average streamflow in PCW.

Figure 3.15 illustrates the spatial distribution of a climate variable such as rainfall over the entire PCW using interpolation between the four grid point estimates of the regional climate generated through GCM experiments. The three points inside PCW show the relative positions of the weather stations where observed climate data used for perturbing the GCM generated values came from. Color-coded lines indicate areas with more or less the same downscaled rate of projected climate change at a certain time period and emission scenario.

Figure 3.15. Illustrative example of a downscaled climate scenario for PCW.

Future Rainfall Trends

In general, rainfall amount is expected to increase in 2020, 2050 and 2080 relative to the observed climatology from 1961-1990 (Table 3.3 and Figure 3.16 and 3.17). For the three time slices, the increase will be less than the GCM results but greater than the projected estimates for Southeast Asia (IPCC 2001). For 2020 the increase of 5.3 mm in the average daily rainfall is projected, 5.5 mm for 2050, and 5.6 mm for 2080. The increase is observed across the various climate scenarios (Figure 3.18–3.20).

The maximum rainfall events per year is also detected to increase in the three future time periods compared to the long term average rainfall records in PCW. Exceptions are three projections: CCCma_A2a, CCCma_B2a 2050, and CCCma_A2a 2080, where the average maximum rainfall event was noted to decrease (Figure 3.16). From the observed average maximum event of 20.6 mm, the increase over most of the projections range between 21.3–26.8 mm.

Table 3.3. GCM and downscaled mean daily rainfall for PCW

	2020				2050				2080			
Scenario	GCM*		Downscaled GM*		GCM*		Downscaled GCM*		GCM*		Downscaled GCM*	
	mm	%	mm	%	mm	%	mm	%	mm	%	mm	%
CCCma_A2a	5.7	11.6	5.5	6.9	5.7	12.5	5.5	7.9	5.5	8.1	5.3	3.4
CCCma_B2a	5.7	12.3	5.3	4.2	5.6	9.9	5.4	5.9	5.7	11.7	5.4	6.1
CSIRO_A2a	5.4	6.1	5.4	5.3	5.4	6.5	5.6	10.3	6	18.1	6	18.4
CSIRO_B2a	5.5	6.9	5.4	6.4	5.6	9.9	5.8	13.5	5.5	8.2	5.7	12.7
HADCM3_A2a	5.2	1.6	5.3	2.9	5.2	1.1	5.2	2.5	5.2	2.4	5.7	12
HADCM3_B2a	5.2	1.6	5.2	1	5.2	1.3	5.3	3.3	5.4	5.2	5.5	8.6
Average	5.4	6.7	5.3	4.5	5.5	6.9	5.5	7.2	5.6	8.9	5.6	10.2
SE Asia IPCC (2001)		2.4		2.4		4.6		4.6		8.5		8.5

*Average daily value; observed average daily value 5.10

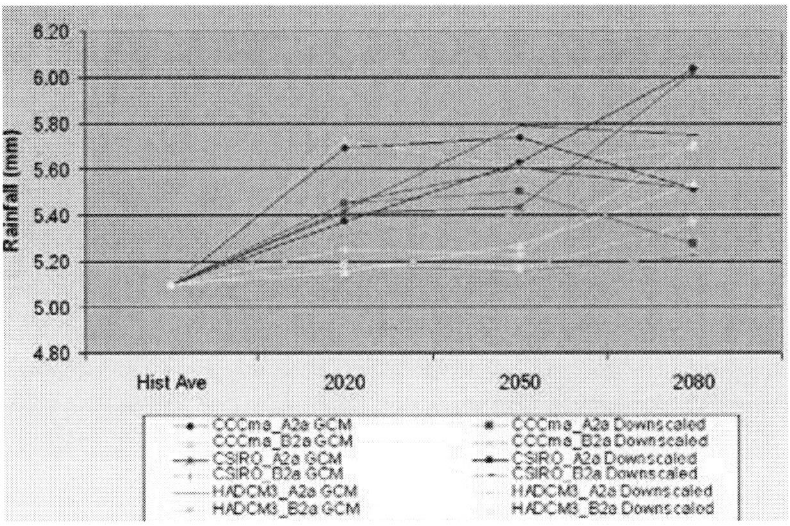

Figure 3.16. GCM-projected and downscaled daily average rainfall for PCW.

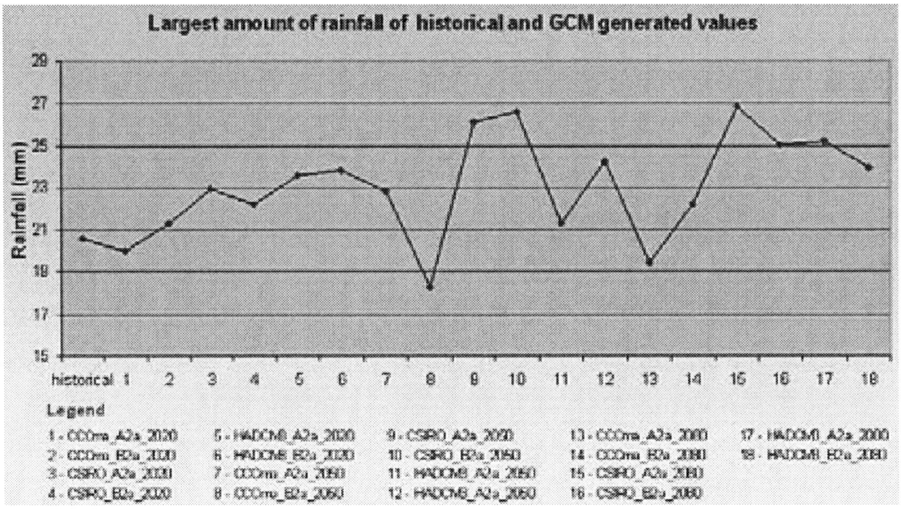

Figure 3.17. Largest amount of rainfall of historical and GCM generated values.

The average daily rainfall per month could generally decrease during the drier months (i.e., December to April) and increase during the wetter months (i.e., May to November) as shown in Figure 3.18–3.20. The average daily rainfall for dry months will fall from the observed average of 0.67–0.58 mm average for all scenarios across all time slices with a decreased rate increasing over time. In contrast, average daily rainfall during wet months will increase from the observed average of 8.3 mm to 9.01 mm average for all scenarios and future time slices.

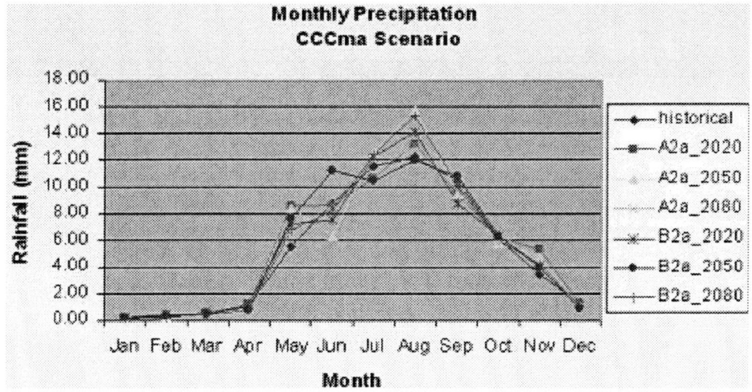

Figure 3.18. Monthly precipitation using CCCma.

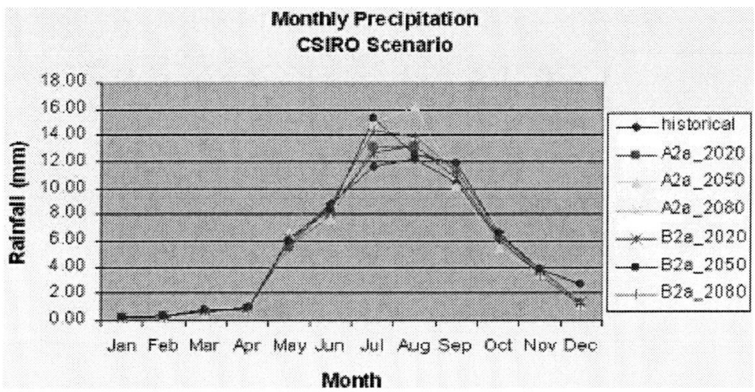

Figure 3.19. Monthly precipitation using CSIRO.

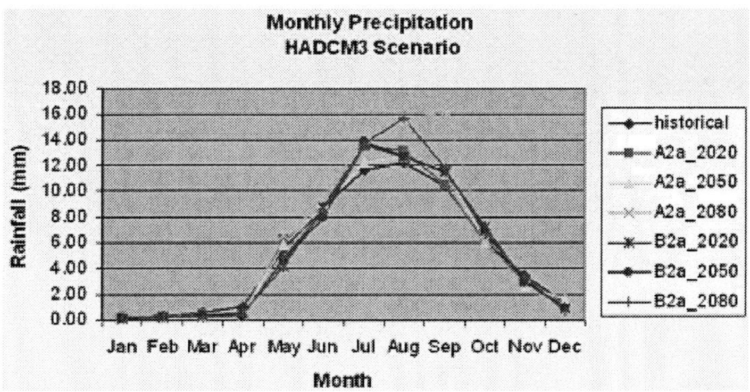

Figure 3.20. Monthly precipitation using HADCM.

Future Temperature Trends

Generally, downscaled temperature is projected to increase in 2020, 2050, and 2080 relative to the observed climatology from 1961-1990 (Table 3.4 and 3.5 and Figure 3.22–

3.31). For the three time slices, temperature is projected to be less than the GCM results. Warming trend is also noted generally across all scenarios and future time periods relative to the observed temperature. These projections for PCW are also higher than the projected estimates for Southeast Asia (IPCC 2001). An exception to this is the tendency of almost half of the projections for minimum daily average temperature to notably cool down in 2020 based on the observed values and thereafter follow a warming path (Figure 3.22).

Average daily maximum temperature is projected to increase over the observed values by 0.7 0C (2.1%) for 2020, 1.23 0C (3.8%) for 2050, and 2.05 0C (6.3%) for 2080 (Table 3.5 and Figure 3.21). For average daily minimum temperature, warming over the observed average values is estimated to be 1.22 0C (3.0%) for 2020, 1.20 0C (5.4%) for 2050 and 2.0 0C (9.0%) for 2080 (Table 3.5).

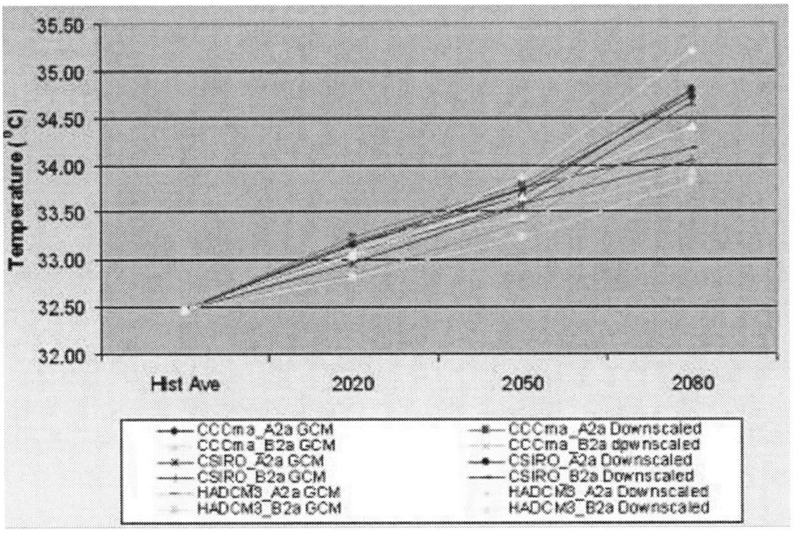

Figure 3.21. GCM and downscaled mean daily maximum temperature scenario for PCW.

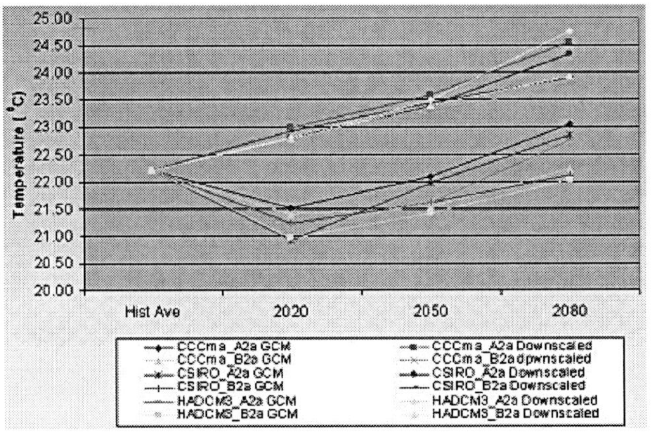

Figure 3.22. GCM and downscaled mean daily minimum temperature scenario for PCW.

Table 3.4. GCM and downscaled mean daily maximum rainfall for PCW

	2020				2050				2080			
Scenario	GCM*		Downscaled GCM*		GCM*		Downscaled GCM*		GCM*		Downscaled GCM*	
	C	%	C	%	C	%	C	%	C	%	C	%
CCCma_A2a	33.2	2.1	33.2	2.3	33.7	3.9	33.8	4.1	34.7	6.9	34.8	7.1
CCCma_B2a	33.1	2	33.2	2.3	33.4	2.9	33.6	3.3	33.9	4.5	33.7	3.7
CSIRO_A2a	33	1.5	33.2	2.1	33.6	3.4	33.7	3.7	34.7	6.7	34.8	7.1
CSIRO_B2a	33.1	1.8	33.2	2.2	33.6	3.4	33.8	3.9	34	4.8	34.2	5.2
HADCM3_A2a	32.8	1.1	33.1	1.8	33.4	2.8	33.9	4.3	34.5	6.1	35.2	8.4
HADCM3_B2a	32.8	1	33	1.7	33.2	2.4	33.7	3.7	33.8	4.1	34.4	5.9
Average	33	1.6	33.2	2.1	33.5	3.1	33.7	3.8	34.3	5.5	34.5	6.3
SE Asia IPCC (2001)		1		1		2		2		2.8		2.8
*Average daily values; observed average daily value 32.48												

Note: SE Asia values are summer average temperature.

Table 3.5. GCM and downscaled mean daily minimum temperature for PCW

	2020				2050				2080			
Scenario	GCM*		Downscaled GCM*		GCM*		Downscaled GCM*		GCM*		Downscaled GCM*	
	C	%	C	%	C	%	C	%	C	%	C	%
CCCma_A2a	21.5	-3.1	23	3.5	22.1	-0.6	23.6	6.1	23.1	3.7	24.6	10.5
CCCma_B2a	21.4	-3.7	23	3.5	21.5	-3	23.3	4.8	22.2	0	23.8	7
CSIRO_A2a	20.9	-5.9	22.8	2.7	22	-1.2	23.4	5.2	22.8	2.8	24.4	9.6
CSIRO_B2a	21.2	-4.4	22.9	3.2	21.6	-2.7	23.5	5.6	22.1	-0.5	23.9	7.6
HADCM3_A2a	21.2	-4.6	22.8	2.7	21.6	-2.7	23.5	5.7	22.8	2.4	24.8	11.4
HADCM3_B2a	21	-5.7	22.8	2.5	21.4	-3.6	23.4	5.3	22	-0.8	23.9	7.7
Average	21.2	-4.6	22.9	3	21.7	-2.3	23.4	5.4	22.5	1.3	24.2	9
SE Asia IPCC (2001)		1.1		1.1		2.3		2.3		3.2		3
*Average daily values; observed value 22.22												

Note: SE Asia values are winter average temperature.

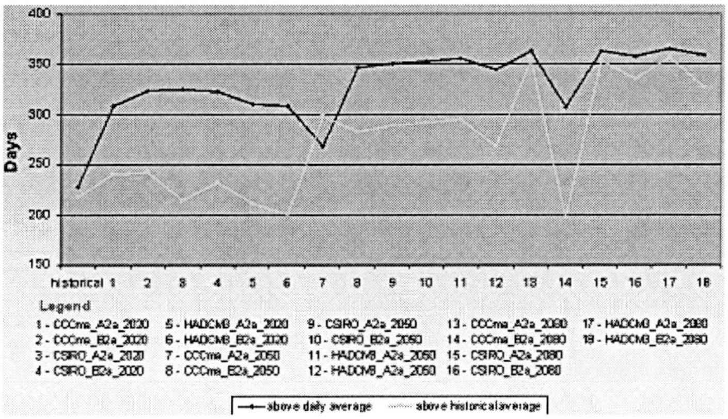

Figure 3.23. Number of days maximum temperature is above daily* and historical average. (*Daily average is the average of daily values by month from 1961-2000).

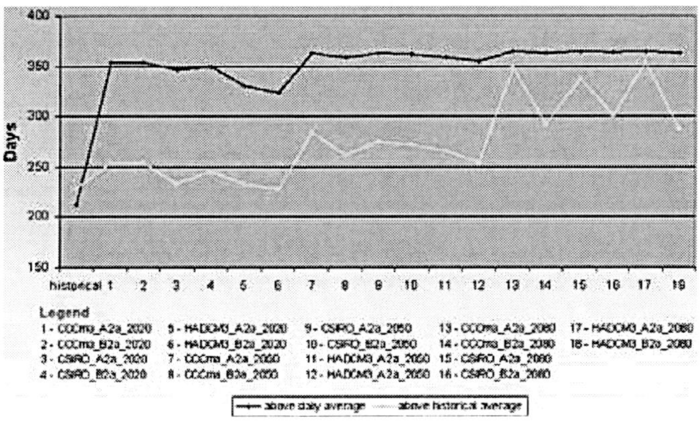

Figure 3.24. Number of days minimum temperature is above daily* and historical average. (*Daily average is the average of daily values by month from 1961-2000).

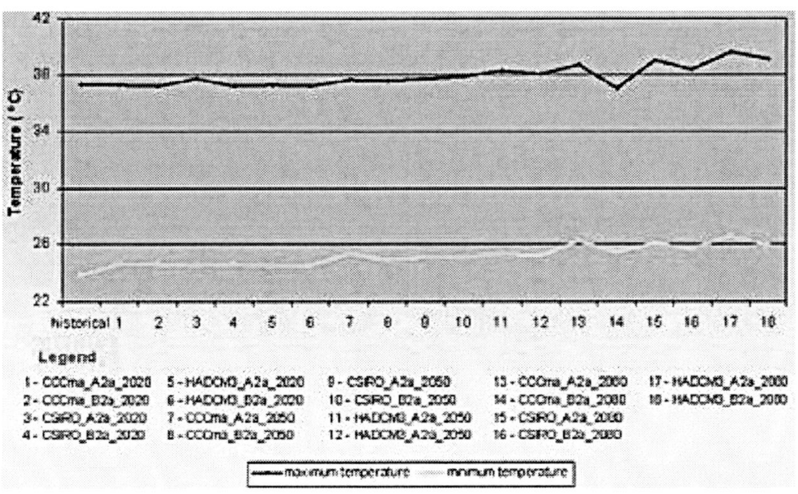

Figure 3.25. Highest maximum and minimum temperature of historical and GCM-generated values.

Table 3.6. Averages of projected average daily maximum temperature for PCW

Month	Observed	Average for all scenarios	Average for 2020 scenarios	Average for 2050 scenarios	Average for 2080 scenarios
January	31.37	32.21	31.69	32.06	32.90
February	32.24	33.45	32.80	33.38	34.17
March	33.49	34.74	34.18	34.74	35.31
April	35.11	36.75	36.06	36.68	37.50
May	35.24	36.61	36.00	36.53	37.31
June	32.71	34.14	33.55	34.13	34.76
July	32.36	33.57	32.93	33.59	34.20
August	30.85	32.24	31.70	32.25	32.77
September	32.05	33.31	32.75	33.27	33.92
October	31.56	33.14	32.54	33.17	33.72
November	32.18	32.54	31.93	32.51	33.19
December	31.57	32.26	31.70	32.17	32.91
Average	32.56	33.75	33.15	33.71	34.39
Dry Average	32.75	33.88	33.28	33.81	34.56
Wet Average	32.42	33.65	33.05	33.64	34.27

Table 3.7. Averages of projected average daily maximum temperature for PCW

Month	Observed	Average for all scenarios	Average for 2020 scenarios	Average for 2050 scenarios	Average for 2080 scenarios
January	20.13	21.33	20.68	21.3	22
February	20.41	21.48	20.93	21.34	22.18
March	21.45	22.65	22.09	22.52	23.33
April	22.97	24.35	23.74	24.25	25.07
May	23.83	24.74	24.06	24.65	25.52
June	23.12	24.63	23.97	24.57	25.34
July	23.61	24.91	24.27	24.79	25.66
August	23.48	24.89	24.24	24.85	25.59
September	23.4	24.75	24.1	24.67	25.49
October	22.27	23.84	23.1	23.82	24.56
November	22.08	22.7	22.05	22.6	23.46
December	21.04	21.91	21.31	21.78	22.62
Average	22.31	23.52	22.88	23.43	24.23
Dry Average	21.2	22.34	21.75	22.24	23.04
Wet Average	23.11	24.35	23.69	24.28	25.09

The warming trend is also seen in the increasing number of days the maximum and the minimum temperatures are above the daily normal values (Figure 3.23 and 3.24). The number of days maximum temperature is above normal values decreased in almost all scenarios for 2020, but is then seen to increase over the normal values in 2050 and 2080. For the number of days the minimum temperature is above the long term average values, the trend is increasing in all three future time slices. Likewise, the highest maximum and minimum temperature values per year is estimated to increase in all three future time periods and across all scenarios compared with the long term observed highest annual maximum and minimum temperatures (Figure 3.25).

The average daily maximum and minimum temperatures per month are estimated to increase during the drier months (i.e., December to April) and during the wetter months (i.e., May to November) as shown in Table 3.6 and 3.7 and Figure 3.26–3.31. The dry months' average daily maximum temperature will increase from the observed average of 32.75 0C to 33.88 0C average for all scenarios across all time slices. In the wet months the average daily maximum and minimum temperature will also increase from the observed average of 21.20 0C to 22.34 0C average for all scenarios and future time slices. During the wet months the average daily minimum and maximum temperatures are also seen to increase from the observed average values, from 23.11–24.35 0C for minimum temperature and from 32.42–33.65 0C for the maximum temperature.

Table 3.8. Range of projected change in climate at PCW using various GCM results

Variable	Range	2020	2050	2080
Rainfall (mm)	Min	0.05 (0.9)	0.13 (2.5)	0.18 (3.5)
	Max	0.35 (6.9)	0.69 (13.5)	0.65 (12.7)
Minimum temperature (0C)	Min	0.56 (2.5)	1.08 (4.9)	1.2 (5.4)
Minimum temperature (0C)	Max	0.76 (3.4)	1.38 (6.2)	2.73 (12.3)
Maximum temperature (0C)	Min	0.57 (1.8)	1.06 (3.3)	1.56 (4.8)
	Max	0.77 (2.4)	1.36 (4.2)	2.54 (7.8)
Note: Values in parenthesis are percent change				

Note: Values in parenthesis are percent change.

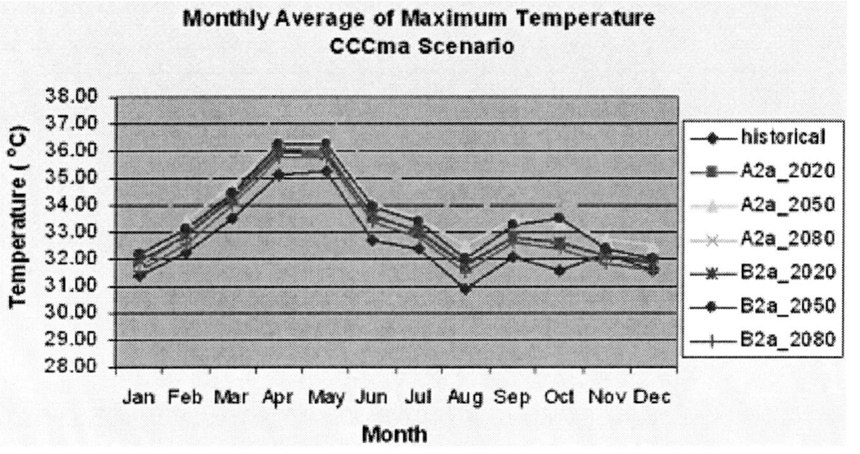

Figure 3.26. Monthly average of maximum temperature using CCCma.

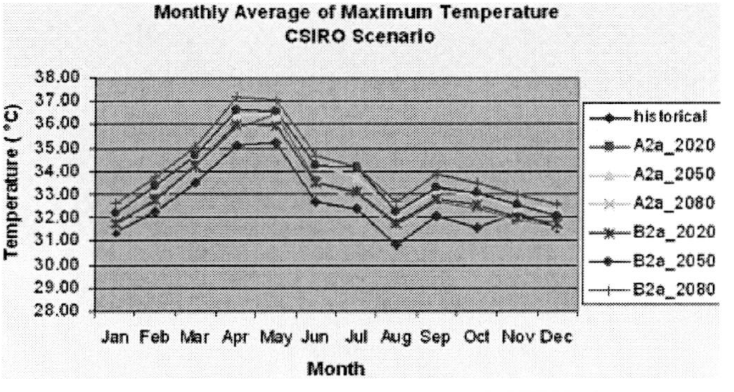

Figure 3.27. Monthly average maximum of temperature using CSIRO.

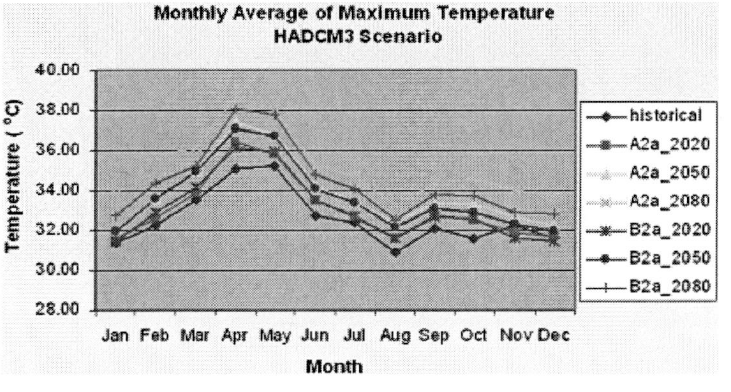

Figure 3.28. Monthly average of maximum temperature using HADCM3.

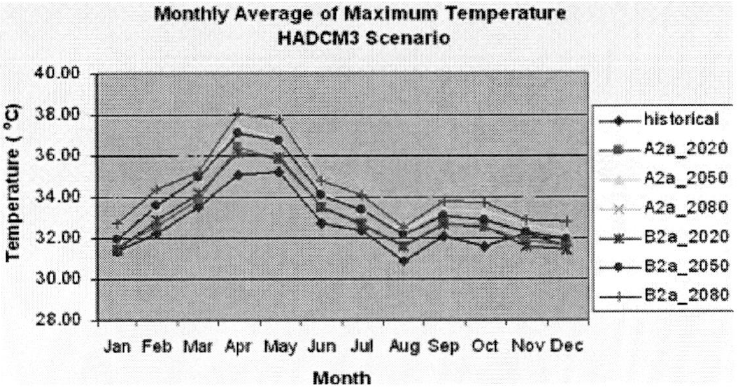

Figure 3.29. Monthly average of maximum temperature using HADCM3.

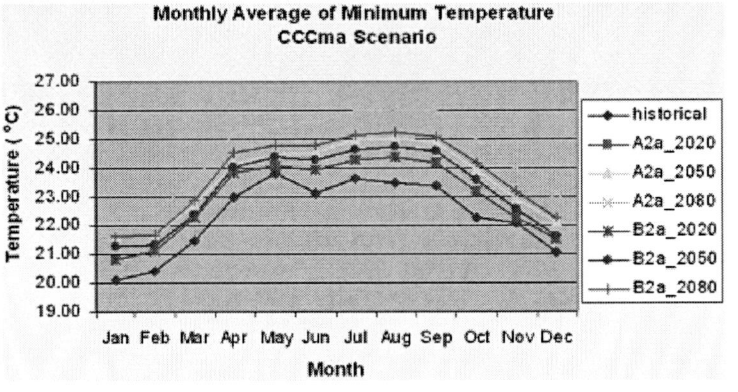

Figure 3.30. Monthly average minimum temperature using CCCma.

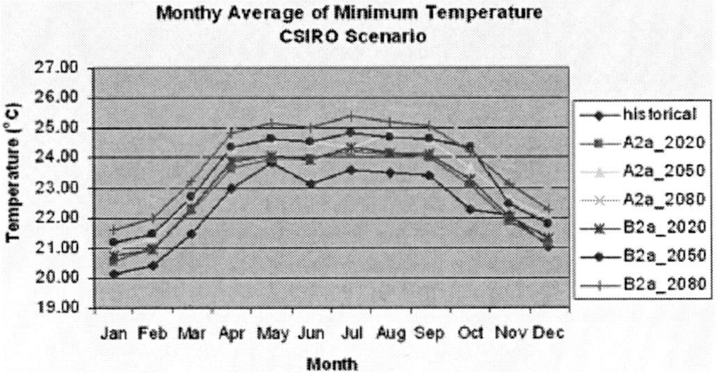

Figure 3.31. Monthly average of minimum temperature using CSIRO.

Chapter 4

IMPACTS AND VULNERABILITY TO CLIMATE CHANGE

Climate change is projected to affect natural and human systems. The Philippines is highly vulnerable to current climate risks as well as future climate change. An average of 20 tropical cyclones enter the Philippine area of responsibility, although around 8-9 will cross any part of the country. These result to a great loss of lives and damage every year. For example, in September 2006, tropical storm Milenyo (international codename: Xangsane) caused the death of 184 people and injured 536 people. It damaged about 500,000 houses with total damages to properties reaching US$ 134 million. Agricultural damage reached US$ 83 million (NDCC 2006). It is expected that climate change will exacerbate existing stresses in the country (PINC 1999).

In this chapter, the potential impacts of climate change in PCW is presented as well as an assessment of vulnerabilities of natural and human systems. In the case of impacts on forest ecosystems, a nationwide study was conducted.

Climate change is expected to affect tropical forest ecosystems of the world. The impacts of climate change on forest ecosystems' vulnerability are a major concern in developing countries like the Philippines. Philippine watersheds are predicted to be adversely affected by climate change effects due to their present critical state and the varying stages of deterioration experienced. Thus, the study assesses the impacts and vulnerability of forest ecosystems in PCW to climate change. During the study a series of workshops, consultation meetings, and focus group discussions were held with local communities, local government units, and stakeholders.

4.1. CLIMATE IMPACTS AND VULNERABILITY USING HOLDRDIGE LIFE ZONES AND GIS

This section presents an analysis on the potential impacts of climate change to forest types in the Phillippines using the Holdridge Life Zones and GIS.

Potential vs Actual Life Zones

Without any anthropogenic influence, simulation of potential forest types at current temperature and precipitation showed that the Philippines would be dominated by the dry tropical, moist tropical and wet tropical forest life zones (Table 4.1). Such a condition must have existed when the Spanish colonizers first set foot in the Philippines in 1521. At that time it is estimated that 90% of the country was covered with lush tropical rainforest. By the year 1900, there were still 70% or 21M ha of forest cover (Garrity et al. 1993, Liu et al. 1993). However, by 1996 there were only 6.1 M ha (20%) of forest remaining (Forest Management Bureau [FMB] 1998).

Table 4.1. Comparison of potential and actual (based on 1993 data) life zones in the Philippines

Life Zone Type	Area Distribution (ha)		Percent Distribution	
	Potential	1993 Life Zone	Potential	1993 Life Zone
Dry Forest	8,763,696.10	1,082,197.20	29.65%	3.66%
Moist Forest	15,149,315.26	3,534,636.30	51.25%	11.96%
Wet Forest	5,646,414.43	2,266,455.20	19.10%	7.67%
TOTAL	29,559,425.79	6,883,288.71	100.00%	23.29%

The average deforestation rate from 1969 to 1973 was 170,000 hectares per year (Forest Development Center 1987). For the past 20 years, it was about 190,000 to 200,000 hectares per year (Revilla 1997). However, in the last few years it was estimated to be in the vicinity of 100,000 ha (Lasco and Pulhin 1998). In the Philippines, the direct and indirect causes of deforestation include shifting cultivation, permanent agriculture, ranching, logging, fuel wood gathering, and charcoal making (Kummer 1990).

Using ArcView the actual forest cover of the Philippines in 1993 is overlaid with the potential life zones predicted by the Holdridge system. As might be expected, all the forest types declined with the highest decline in dry forests and the least decline in wet forests (Table 4.1).

4.2. Effects of the Synthetic Climate Change Scenarios to Vegetative Cover

Scenario 1. A 25% increase in precipitation will lead to the following changes:

- Total loss of all dry forests even at the lowest temperature change of 1 0C. This is expected considering the increase in available water;
- A 30.5% increase in moist forest across all temperatures;
- A negligible (0.1%) increase in wet forest.

It is noteworthy that temperature increase has negligible drying effects on life zones in the Philippines. Normally, slight increases in temperature (even by 0.5 0C) can already cause El Niño Phenomenon in some portions of the country. This is probably because all parts of the country already fall within the tropical belt under the Holdridge system (> 24 0C) and the precipitation estimates for the different scenarios are quite large, negating any effect on the increase in temperature.

Scenario 2. A 50% increase in precipitation will lead to the following changes:

- Total loss of all dry forests even at the lowest temperature change. This is expected considering the increase in available water;
- A 47% decline in moist forests across all temperatures;
- An increase in rainforests from nothing under current conditions to 365,000 ha under a 1 0C increase in temperature. As temperature increases, there is a slight decline in rainforest area;
- A 106% increase in wet forest cover as a result of greater precipitation.

It is assumed in this study that the total forested area remains fixed in the analysis. In other words, it is assumed that non-forested areas could never revert back to forests because of human influence (e.g., agriculture, settlements).

Scenario 3. A 100% increase in precipitation will lead to the following changes in the life zone pattern in the Philippines:

- Total loss of all dry forests (> 1 M ha);
- A 50% decline in moist forests area;
- A significant rise in area of rain forests from zero under current conditions to more than 2 M ha;
- A 32% increase in wet forests, about one-third of the increase under Scenario 2.

Just like the previous scenarios, the effect of temperature change is minimal due to very high precipitation increases. Overall, the simulation study shows that increases (25%, 50% and 100%) in precipitation and temperatures would result to a re-distribution of forest types as classified by Holdridge. The dry forests are the most vulnerable. They could be totally wiped out even under a 25% increase in precipitation. Moist forests are also vulnerable especially under higher precipitation increase. On the positive side, there will be a significant increase in rain forest types as precipitation level increases.

The results of the study are generally consistent with the findings of the IPCC Second and Third Assessment Reports (IPCC 1996 and IPCC 2001, respectively). At the eco-physiological level, while the net effect of climate change on net primary production (NPP) is not yet clear, there is generally positive correlation between NPP and temperature. However, if rainfall is not sufficient, water stress could be a problem. The TAR points out that tree growth is negatively correlated to annual mean (night) temperature, however, lowland humid tropics are already close to optimum temperature so that there will be no effect on climate change.

Global vegetation models (e.g., BIOME, MAPSS, IMAGE) do not agree on whether tropical forests will increase or decrease (IPCC 2001, IPCC 1996). But any major shift in rainfall pattern will affect distribution of vegetation types. Under enhanced CO2 concentration, tropical evergreen broadleaf forests could readily establish after deforestation. Shifts in rainfall patterns could increase conversion of forests to agricultural land by increasing migration from areas affected by drought, erosion, and so on. Productivity will increase or decrease depending on the amount of rainfall.

4.3. Impacts of Current Climate Extremes and Variability in PCW: Stakeholders' Perspectives

The study employs a combination of data collection methods: secondary data gathering, focus group discussions, workshops, consultation meetings, and direct field observation. Available secondary information on the biophysical aspect of the watershed are gathered from relevant agencies. These include data on land use, climate, soil, and topography.

Focus group discussions (FGDs) are done to determine the impacts and vulnerability of forest ecosystems to climate change and, at the same time, validate present land use of the watershed. Participants of the FGDs include stakeholders, LGUs, non-government organizations (NGOs), and local communities. Qualitatively, the degree of present vulnerability of the forest ecosystems was assessed by aggregating and analyzing the results of FGDs.

The following impacts of climate variability and extremes in the forest ecosystems in PCW were identified by the local communities, stakeholders, LGUs, and people's organizations during workshops, consultations, FGDs, and interviews.

4.3.1. Lowland Farms

Early and Late Onset of Rainy Season

The cropping pattern of lowland farms in Carranglan is altered when rainy season begins earlier than normal. In Pantabangan, on the positive side, early onset of rainy season results to earlier planting of palay and other agricultural crops. Participants from Pantabangan also noted that during this season there is an increase in the production of energy and irrigation. Thus, lowland farms in Carranglan are considered to be strongly vulnerable to impacts of early onset of rainy season while lowland farms in Pantabangan are moderately vulnerable.

Similarly, late onset of rainy season alters the cropping pattern of lowland farms in Carranglan. In Pantabangan rainfed farms in lowland areas are the most affected by late onset of rainy season, since they depend heavily on frequency of rain during the year. Moreover, decreases in the production of palay, other agricultural crops, and irrigation water are also observed. Thus, lowland farms in Pantabangan are strongly vulnerable to late onset of rainy season.

La Niña

La Niña causes flooding and excessive crop damages to lowland farms in Carranglan. Consequently, these result to occurrences of excessive soil erosion and revenue losses. On the other hand, occurrences of flash flooding in Pantabangan lowland farms are usually observed.

These result to less soil fertility, erosion, siltation, and low agricultural production. In terms of vulnerability to La Niña, lowland farms in Carranglan are more vulnerable compared with farms in Pantabangan.

El Niño

El Nino event results to stagnation of the growth of the crops and most often leads to death of crops in Carranglan lowland farms. As a consequence, farmers have low crop production, hence, low revenue. Likewise, occurrences of fire that can destroy lowland farms are likely to happen during El Niño episodes. This is due to the presence of highly combustible agricultural wastes that can easily ignite. Similarly, productions of agricultural products from Pantabangan's lowland farms are affected by the El Niño event. Irrigation water is also scarce during El Niño events since only farms near the source are being serviced with irrigation water. Thus, lowland farms in both Pantabangan and Carranglan areas are very strongly vulnerable to El Niño.

High Temperature/Summer Season

High temperature/summer season reduces the productivity of the lowland farms in Carranglan and Pantabangan, resulting to decreased revenue. To increase income from farming, farmers use more inputs which result to increased farm costs. In terms of degree of vulnerability, lowland farms in Pantabangan are more vulnerable compared with lowland farms in Carranglan.

Rainy Season

Rainy season has a favorable effect on lowland farms of Pantabangan and Carranglan. During this season, farmers incur lower costs and gain more profits in producing agricultural products. Thus, lowland farms in Pantabangan are only moderately vulnerable to rainy season.

4.3.2. Upland Farms

Early and Late Onset of Rainy Season

Early onset of rainy season results to early cultivation of upland farms and expansion of kaingin areas in Pantabangan. According to farmers, there is relatively high agricultural production during such time resulting to higher income from farming.

Late onset of rainy season, on the other hand, results to further destruction of forests in the Pantabangan area because farmers have to resort to other sources of livelihood to feed their families. Farmers experience low farm production when rain comes late. Farm produce becomes insufficient to meet family needs.

Thus, in terms of vulnerability, upland farms in Pantabangan are moderately affected by early onset of rainy season but are very vulnerable to late onset of rainy season. For farmers in Carranglan, early or late onset of rainy season has adverse impacts on the cropping pattern. This climate variability has not been observed by farmers from Ma. Aurora because the climatic condition in the area, according to participants, is more or less the same all throughout the year.

La Niña

Too much rain resulting from occurrence of La Niña affects crops. Farmers mention that flooding and excessive soil erosion occur during this type of climate extreme. Soils in the upland farms lose its fertility and crops are damaged as consequence. This observation is true for upland farms of Carranglan and Ma. Aurora. In Pantabangan, however, upland farms are not very much affected by La Niña. According to participants, there is less erosion occurring in their upland farms despite the pouring of more rains. Among the three municipalities, Carranglan area is the most vulnerable to La Niña. While Pantabangan is moderately vulnerable, Ma. Aurora is slightly vulnerable.

El Niño

Although characteristics of La Niña and El Niño episodes are very much different, both of them have similar effects on the crops. As what the participants from the Carranglan and Pantabangan areas mentioned, both La Niña and El Niño are damaging to crops. The unavailability of moisture during El Niño episodes causes the crops to experience water stress that inhibits the growth of the plants. For flowering crops, El Niño limits or at worst hinders the production of flowers of the plants.

This obviously results to non-productivity of the crops and loss of farmer income. For some farmers El Niño episodes result to loss of opportunity to plant since rain is an essential element in the early stages of life of the plants/crops. Furthermore, the participants note that extreme dryness increases the possibility of the occurrence of fire in the area resulting to burning and/or damaging of the upland farms.

Aside from its effect on the growth of the plants/crops, the absence of rain results to extreme dryness of the soil causing soil organisms to die. Soil microorganisms are important components of the soil. They mix the plant and animal residues into the soil creating humus, an essential component to attain good soil physical and chemical conditions (Brady 1984).

No impact has been noted in the Ma. Aurora area because, according to the participants, they barely experience El Niño event in the area. In terms of the degree of vulnerability of the upland farms in Pantabangan and Carranglan, the participants noted that the upland farms are very much vulnerable to El Niño.

High Temperature/Summer Season

Impacts of high temperature/summer season are quite similar to the observed impacts of El Niño. According to the participants from Carranglan and Pantabangan, this climate variability causes damage to crops. Likewise, during this season, upland farms are observed to be prone to fire. On the positive note, however, high temperature/summer season prevents attacks of fungi because there is not much moisture present. Fungi thrive well in moist areas.

On the other hand, upland farms in Ma. Aurora do not experience any impact caused by high temperature/summer season because there is rain all year round. The degree of vulnerability varies among areas. For instance in Carranglan, the participants reported that the upland farms are not very vulnerable to high temperature/summer season while Pantabangan is very vulnerable to the mentioned climate variability.

Rainy Season

Characterized by an abundant supply of water, this climate variability is found to be favorable to the growth of crops in upland farms in Carranglan. However, with too much moisture available during this season, risk of attack of fungal infestation is very high. In Pantabangan rainy season results to expansion of kaingin farms resulting to reduction of forest cover, occurrence of soil erosion and decrease in soil fertility. On the positive note, the participants mention that rainy season results to increased agricultural production. The degree of vulnerability of upland farms in Carranglan and Pantabangan to rainy season is not very profound. According to participants, upland farms in the mentioned areas are only moderately vulnerable to rainy season.

4.3.3. Tree Plantations

Early or Late Onset of Rainy Season

Early onset of the rainy season has positive effect on the trees in the plantation areas of Pantabangan, Carranglan, and Ma. Aurora. Early onset of rain results to fast growth rate of trees. As a consequence, the areas covered by PCW are moderately affected by early onset of the rainy season. Late onset of rainy season, on the other hand, results to the death of seedlings in the Carranglan area, thus, decreasing survival rate of seedlings planted. In Pantabangan grown trees in the plantation areas dry up whenever rain comes in late. During this time the participants mention that planting of trees is not at all possible. In Ma. Aurora late onset of rainy season results to moderate growth of trees or decreased productivity of tree plantation areas.

Degree of vunerability in the plantation areas of Pantabangan, Carranglan and Ma. Aurora varies with the type of climate variability. For instance, tree plantations are moderately or slightly vulnerable to early onset of the rainy season while these same areas are strongly vulnerable to late onset of the rainy season.

La Niña

La Niña has no pronounced impact on the tree plantations in Ma. Aurora. But on the other hand, tree plantation areas in Carranglan and Pantabangan are affected by the occurrence of La Niña. Because of too much rain, tree plantations are more prone to pests and diseases resulting to decay of trees. In some instances, growth of trees is hampered during La Niña events. Likewise, this climate variability enhances occurrence of soil erosion in the area because the soil particles become loose due to excessive rainfall. In terms of degree of vulnerability of the plantation areas, it was noted by the participants from Carranglan that tree plantation areas in their municipality are strongly vulnerable to La Niña events.

El Niño

Impacts of El Niño episodes to tree plantation areas in Carranglan include: (1) increased vulnerability to fire; (2) increased mortality due to drought; (3) increased cost of maintenance; (4) decreased activity of the soil organisms; and (5) stunted/slow growth of trees. In Pantabangan, however, participants mention that the growth of trees are observed to be slow. In Ma. Aurora El Niño has no pronounced impact because, as mentioned earlier, rain is evenly distributed all throughout the year. Among the municipalities covered by PCW, tree plantation areas in Carranglan are observed to be very strongly vulnerable to El Niño events.

High Temperature/Summer Season

The degree of impacts of high temperature/summer season is less compared with that of El Niño in the Pantabangan and Carranglan areas. Since tree plantation areas get enough rain in Ma. Aurora, no impact of El Niño has been recorded. In terms of degree of vulnerability of the tree plantations to El Niño event, participants mention that the ecosystem is moderately vulnerable to the mentioned climate variability.

Rainy Season

The rainy season is found to be favorable to tree plantation areas in PCW. As mentioned by participants, such season is the best time to plant trees because there is sufficient water supply. It also promotes good growth to already established tree plantations. Furthermore, a participant mentioned that rainy season is found to promote less maintenance cost in the tree plantation areas. Degree of vulnerability of the plantation areas is mild only, thus no adaptation measure is undertaken in the area.

4.3.4. Grasslands

Early Onset of the Rainy Season

Experiences show that early onset of rainy season protects grassland areas from fire. This observation is also true in Pantabangan and Carranglan grasslands. Participants also mentioned that grasslands are less prone to fire during this season because grasses are moist. In addition, Pantabangan grasslands are observed to regenerate early during this season. However, early onset of rainy season promotes occurrence of soil erosion. Thus, while grasslands in Pantabangan and Carranglan are moderately vulnerable to early onset of rainy season, grasslands in Ma. Aurora are not vulnerable to this climate variability.

Late Onset of Rainy Season

Late onset of rainy season results to drying up of grasses and accumulation of fuel in grasslands. Consequently, Pantabangan and Carranglan grasslands become more prone to fire during this season. Moreover, in Pantabangan soil becomes too dry resulting to cracking of the soil.

La Niña

Characterized by too much rainfall, La Niña episodes cause soil erosion in the grasslands of Pantabangan and Carranglan. Furthermore, in Carranglan La Niña causes flooding and death of soil organisms resulting to decreased soil fertility. Aside from soil organisms, the mortality of birds, rats, and other predator species were observed during La Niña event. Hence, grasslands in Pantabangan and Carranglan are moderately vulnerable to La Niña.

El Niño

Extreme dryness caused by El Niño results to occurrence of grassland fires in Pantabangan and Carranglan. El Niño episodes also reduce the amount of forage resulting to lower cattle production. In Carranglan there is also decreased soil fertility resulting from a decrease in soil organism population. As mentioned before, soil organisms are important components of soil to maintain its fertility. However, El Niño also eradicates unwanted microorganisms and is favorable in some ways. Therefore, grassland areas in Pantabangan and Carranglan areas are strongly vulnerable to El Niño.

High Temperature/Summer Season

The effects of high temperature/summer season in Carranglan grasslands are similar to the effects of El Niño. However, observed impacts were in lesser degree during summer season/high temperature compared with the impacts of El Niño. In Pantabangan grasses are observed to die and soils become more compacted either during the summer season or when temperature is extremely high. In terms of degree of vulnerability, grasslands in Pantabangan and Carranglan are strongly vulnerable to high temperature/summer season.

Rainy Season

Rainy season is found to be favorable to grassland areas in Carranglan. However, in Pantabangan rainy season has both positive and negative effects. According to participants, negative effects of rainy season are erosion, siltation, and flash floods while a positive effect includes increased livestock production. Generally, grasslands in Pantabangan and Carranglan are mildly vulnerable to rainy season.

4.3.5. Natural Forests

Early Onset of the Rainy Season

Early onset of the rainy season affects the flowering of the trees in natural forests of Carranglan. While no concrete effect was mentioned, it is assumed that early onset of rainy season causes early flowering of trees. In Pantabangan, however, presence of water in the springs is observed during this climate variability. Likewise, the participants mentioned that early onset of rain promotes early composting of the soil resulting to higher soil nutrition. Natural forests in Carranglan are moderately vulnerable to the early onset of rainy season while natural forests in Pantabangan are strongly vulnerable.

Late Onset of the Rainy Season

When there is late onset of rainy season, it is observed that trees in the natural forests of Pantabangan dry up. As a consequence, wildlife present in the area migrate to other thickly forested areas. In terms of vulnerability, natural forests of Pantabangan are moderately vulnerable to impacts of late onset of rainy season.

La Niña

Impacts of La Niña vary depending on the location of natural forests. For instance, while La Niña affects cross pollination of trees in the natural forests of Carranglan, Pantabangan forests are destroyed and soil erosion occurs during La Niña events. In natural forests of Ma. Aurora, occurrence of soil erosion is the common impact of La Niña. In terms of degree of vulnerability, participants from Pantabangan note that natural forests in their area are moderately affected by La Niña while participants from Ma. Aurora mention that natural forests in their place are very much vulnerable to La Niña.

El Niño

Compared with La Niña, impacts of El Niño are more severe. For instance, natural forests in Carangglan are in greater risk of forest fire during El Niño event. Moreover, extraction of forest products in the natural forests is more prevalent during El Niño because climate is very dry. Access roads to natural forests are passable and safe, hence, very conducive to transporting harvested forest products. Consequently, reduction of forest resources is observed during El Niño. Another impact of El Niño to natural forests is the reduction of available water needed for the growth of trees. As a result, there is less forest productivity observed during El Niño event. In Pantabangan, however, El Niño has no significant impact in most parts of the natural forests, although there are portions of the mentioned ecosystem that are affected. Hence, natural forests of Carranglan and Pantabangan are moderately vulnerable to El Niño event.

High Temperature/Summer Season

High temperature/summer season affects only portions of the natural forests in Pantabangan while it has no significant effect on the natural forests of Carranglan and Ma. Aurora. In terms of degree of vulnerability, natural forests of Pantabangan are moderately prone to high temperature/summer season.

Rainy Season

Similar to the noted effects on natural forests whenever there is high temperature/summer season, rainy season brings very small impact on the natural forests because there are only small portions of natural forests that are affected. Moreover, the mentioned ecosystem is only moderately vulnerable to rainy season.

4.3.6. Soil and Water

Early Onset of the Rainy Season

Early onset of the rainy season results to abundant supply of water in all the three areas. In the soil there is increased decomposition rate resulting to increased soil fertility. Soil and water in the three areas are moderately vulnerable to early onset of the rainy season.

Late Onset of Rainy Season

Late onset of rainy season results to hardening of soil in Carranglan while in Pantabangan the consequences include: decrease in elevation of water level in dam, decrease in agriculture, livestock, and potable water supply, and increase in water demand. In Ma. Aurora late onset of the rainy season results to decreased stream/spring flow. The degree of vulnerability of the three areas to late onset of rainy season varies. For instance, while Carranglan and Ma.

Aurora are moderately vulnerable to late onset of rainy season, Pantabangan is strongly vulnerable .

La Niña

The impacts of La Niña in soil and water resources of Pantabangan, Carranglan, and Ma. Aurora are the same. These include: (1) replenishment of water in the reservoir; (2) occurrence of excessive soil erosion resulting to siltation in the water resources; (3) flooding; and (4) loss of soil fertility. In terms of degree of vulnerability, however, it was noted by participants that among the three areas, soil and water resources of Ma. Aurora were less likely to be vulnerable compared with those in Carranglan and Pantabangan.

El Niño

The absence of enough water results to the lowering of ground water level in all three areas covered by PCW. Moreover, according to the participants from Carranglan, the soil in their area cracks during El Niño event. Notable observations in Pantabangan, other than lowering of the water table, include: (1) acidity of the soil decreases; (2) soil nutrients are reduced; and (3) areas that are usually underwater appear during drought days resulting to availability of additional areas for crop production. Among the three areas, participants noted that soil and water resources of Ma. Aurora are moderately vulnerable while soil and water resources of Carranglan and Pantabangan are very strongly vulnerable to El Niño.

High Temperature/Summer Season

According to participants, impacts of high temperature/summer season to soil and water resources of Pantabangan are the milder versions of the effects of El Niño in the same resources. Participants in Carranglan noted that there is rapid evaporation and disturbance of activities of soil microorganisms. Pantabangan's soil and water resources are strongly vulnerable to high temperature/summer season while the same resources of Carranglan are very strongly vulnerable to this season.

Rainy Season

In general, rainy season is favorable to soil and water resources of PCW. Specifically, it results to abundance of water supply both for domestic and irrigation purposes. It is also favorable for land preparation needed for crop production. Thus, rainy season only moderately affects the soil and water resources of Pantabangan.

4.4. Current Impacts and Vulnerabilities

Floods and Droughts

There are two major concerns related to water resources in the watershed as influenced by climate change and variability and extreme events, floods, and droughts. These concerns become more vital in the light of downstream communities depending on the PCW for their irrigation needs and protection against floods. As defined by the NIA, the number of months when floods and droughts were of particular concern is determined based on the level of water in the reservoir shown below (Table 4.2).

Table 4.2. Criteria for determing number of months when floods and droughts were significant

For Flood	For Water Shortage
Low – water level < 190 m	Low – water level > 190 m
Mod – water level 191-215 m	Mod – water level 176-189 m
High – water level > 215 m	High – water level < 176 m

Table 4.3 shows that most of the time there seems to be abundant water in the reservoir as indicated by the large number of months when flood is a moderate concern and by the small number of months when water shortage is of low concern. However, it is worth noting that there is a tendency for the number of months when water shortage is of high concern to increase, indicating the imminent risk of water shortage to lowland farmers. This is illustrated by the sudden drop of rice harvest by about 0.3 tons ha-1 based on average production during the dry season (Table 4.4).

Hydroelectric power generation, though not the primary use of the multipurpose reservoir supported by PCW, plays a critical role in the overall power supply mix of the country. It is shown in Figure 4.1 that from a maximum of more than 300,000 Mwh in the early 80s the gross power generated in the reservoir dropped to almost zero in the late 90s. This is mainly attributed to the decreased capacity of the reservoir to stor.e enough water as a result of excessive siltation (Table 4.5).

Table 4.3. Number of months when water level in Pantabangan Dam was critical

Year	Flood Concern			Water Shortage Concern		
	Low	Mod	High	Low	Mod	High
1977		9		9		
1978		8	4	12		
1979		12		12		
1980	5	6		6	6	
1981	3	9		9	3	
1982	4	8		8	4	
1983	7	5			7	5
1984	8	4		4	5	3
1985	5	7		7	5	
1986	4	8		8	4	

1987	9	3		3	9	
1988	10	2		2	8	2
1989	8	4		4	6	2
1990	5	7		7	5	
1991		12		12		
1992	5	7		7	5	
1993	7	5		5	4	3
1994	4	8		8	4	
1995	8	4		4	4	4
1996	9	3		3	8	1
1997	11	1		1	9	2
1998	9	3		3	3	6
Total	121	135	4	134	99	28

Table 4.4. Irrigation performance of UPRIIS, PCW from 1990-2001

Year	Service Area (ha)	Irrigated/Planted Area (ha)				Yield (cavans ha-1)	
		Wet Season		Dry Season		Wet Season	Dry Season
		Program	Actual	Program	Actual		
1990	103,285	91,045	84,813	78,285	77,402	77	78.25
1991	103,285	87,700	83,701	83,603	77,568	78	79.25
1992	103,285	88,843	86,058	80,676	78,441	79.25	81
1993	103,285	90,145	83,252	75,549	75,655	79.75	82
1994	103,285	92,447	85,674	75,792	77,800	80	81.75
1995	103,285	89,233	84,566	71,265	69,975	80.5	81.75
1996	103,285	89,008	81,850	48,694	56,501	81	82.5
1997	103,285	89,001	83,805	54,259	55,501	81.75	83.5
1998	103,285	87,144	74,693	50,575	48,484	81.25	74.25
1999	103,285	87,446	82,857	70,461	74,689	82.5	85.25
2000	103,285	90,293	81,392	79,680	78,380	76	80.25
2001	103,285	89,464	85,188	80,292	75,075	-	-
Average	103,285	89,314	83,154	70,761	70,456	79.73	80.89

Note: 1 cavan of unmilled rice is approximately 50 kg.

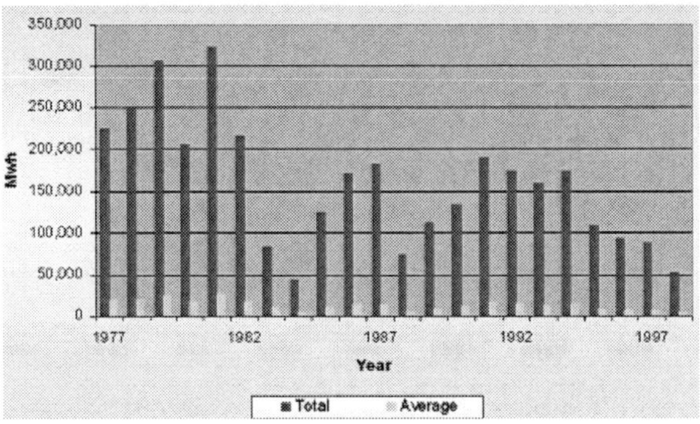

Figure 4.1. Gross power generation (Mwh) at UPRIIS, PCW.

Table 4.5. Sedimentation rates in Pantabangan Reservoir, PCW

Year of Survey	No. of Monsoon After Impoundment	Volume of Sediment (MCM)	Sedimentati on/Year (MCM yr-1)	Total Capacity Lost (%)	Estimated Sediment Yield1 (t ha-1)
1985	12	84.74	7.06	5.5	114.7
1989	15	105.25	6.84	6.6	107.2
1995	21	147.42	7.02	7.6	110.26
2000	26	197.6	7.6	10.2	123.5
Designed for : CY 1989 Survey					
Sedimentation Rate = 1.3 MCM/year3			Sedimentation Rate = 6.84 MCM/year		
Reservoir Life = 100 years			Projected Reservoir Life2 = over 100 years		

Note: 1. Estimated catchments sediment yield is based on the following:

a. Sediment Density = 1,320 kg m3

b. Sediment Trapping Efficiency is 98%

c. Drainage/Catchments Areas is 82,900 ha

2. The Projected Reservoir Life is calculated using the Empirical-Area-Reduction Method
3. Estimated Sedimentation Rate obtained during the Feasibility of the Dam. Based on the actual sediment pool/storage of the dam, designed rate is about 7.50 mcm yr -1.

The siltation problem in the reservoir could be related to the increasing incidence of monsoonal rains. It is, however, common knowledge that the watershed has been suffering from severe degradation of the forests, chronic grass fires and extensive cultivation of the upland areas. For so many years, reforestation efforts involving the people in various modes together with physical protection measures constitute the core of strategies to reduce the magnitude of siltation problems in the reservoir. Unfortunately, efforts to sustain the early gains from reforestation initiatives failed as funds from donor agencies ran out. In the late 90s water transfer from an adjoining basin was initiated to augment the receding water storage in the reservoir while minimal efforts in watershed protection from further degradation and restoration of denuded areas continue.

Lowland Rice Production

The sensitivity of the service area to changes in water resources coming from PCW reservoir is evaluated in terms of drought and flood susceptibility. The proneness of the various districts of the irrigation service area is determined through FGDs and interviews of key informants from the farmers groups, local government officials, and the NIA.

Occurrences of water shortage and drought in the service area in the past 25 years are identified by respondents and listed in Table 4.6. The respondents recall water shortages and floods that occurred in 1980s and 1990s and notes specifically the El Niño episode in 1998 that brought extreme water shortage in the service area. They attribute the incidence of water shortage and flood to various reasons as listed in Table 4.7.

The respondents also note that the location of rice lands in the service area is one of the major reasons why the farmers experience water shortage. Municipalities that lie at the tail of the service area are the most affected especially during dry season. Similarly, the respondents attribute occurrences of floods to location and natural calamities such as typhoon and heavy rainfall. Figure 4.2 and Figure 4.3 show that the areas that suffer the most from droughts and floods are those located along the fringes of the service areas, mainly owing to the topographical and topological limitations. These areas invariably experience decline in rice production levels from 20% to 100% (of 90–100 cavans ha-1) and escalation in production cost by at least 25% during periods of drought (Table 4.8).

Table 4.6.Water shortage and flood occurrences identified by respondents from key informant interviews and FGDs

Year	District 1	District 2	District 3	District 4
1980s	Extreme water shortage	Extreme water shortage	Extreme water shortage	Extreme water shortage
1980s-present	Flood during rainy season	Flood during rainy season	Flood during rainy season	Flood during rainy season
1997	Extreme flood			
1998	El Niño	El Niño	El Niño	El Niño
2000-present	Water shortage and floods	Water shortage and floods	Water shortage and floods	Water shortage and floods

During floods most of the vulnerable areas suffer between 40-100% (of 60-90 cavans ha-1) loss in production. Incidence of water-borne diseases, damages to irrigation structures, properties, and livestock were also identified by the farmers and other key informants as common problems associated with floods.

It is interesting to note that based on the accounts of many farmers during the FGDs and interviews, water shortage and floods occur yearly in the most vulnerable areas. This highlights a common discrepancy between the appreciation by the farmers and by government agencies of what constitutes a problem or risk to the farmers. This is attributable to the inability of the methods used in the formal or official assessment to measure risks and hazards caused by natural events.

Table 4.7. List of reasons of floods commonly cited by the respondents

Description	District 1	District 2	District 3	District 4
Water Shortage	Occurrences of El Niño	Occurrences of El Niño	Occurrences of El Niño	Occurrences of El Niño
	Poor irrigation infrastructures	Poorly maintained irrigation canals	Poor irrigation infrastructures	Poor irrigation infrastructures
	Distance from the dam and low water level in the reservoir	Improper use of irrigation water	Distance from PCW	Distance from PCW
	Greediness of other farmers and irrigators			
Flood	Natural calamities (heavy rainfall) and deforestation	Natural calamities (heavy rainfall) and deforestation	Natural calamities (heavy rainfall) and deforestation	Natural calamities (heavy rainfall) and deforestation
	If the farmers in the upper service area are not using the water, they let the water flow down to the low-lying service areas	Poor irrigation canals and drainages, and poor construction of water controls	Poor irrigation canals and drainages, and poor construction of water controls	Poor irrigation canals and drainages, and poor construction of water controls
	Location (located in low-lying area)	Location (located in low-lying area)	Location (located in low-lying area)	Location (located in low-lying area)

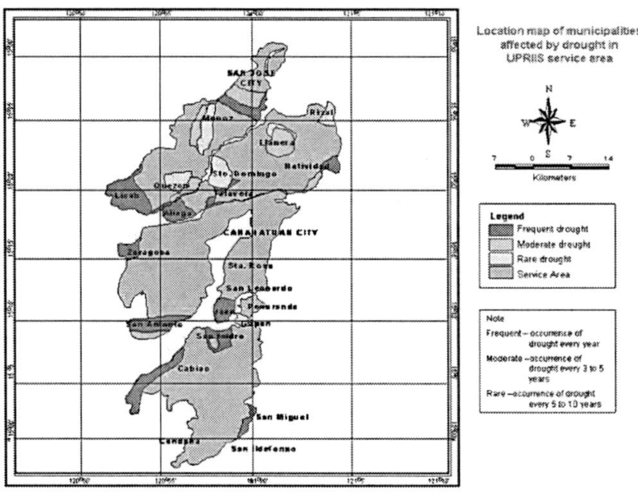

Figure 4.2. Current vulnerability to droughts of the various service area districts of UPRIIS, PCW.

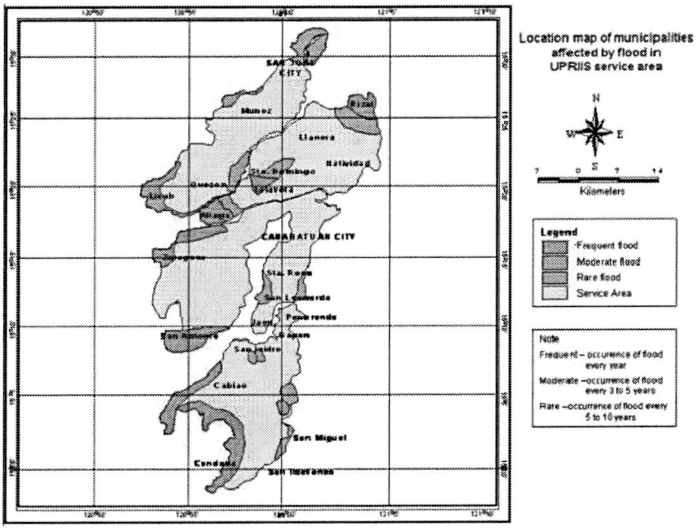

Figure 4.3. Current vulnerability to floods of the various service area districts of UPRIIS, PCW.

Table 4.8. Current vulnerability to floods and droughts of the different service area districts of UPRIIS, PCW

Description	District 1 (Licab and Quezon)	District 2 (Talavera and Aliaga)	District 3 (Jaen)	District 4 (San Miguel, Bulacan and Candaba, Pampanga)
Water Shortage				
Impacts	Decreased productivity by at least 20% High production cost for agricultural products	Decreased productivity by at least 70% (from 100 cavans per hectare to 15-30 cavans ha-1) High production cost for agricultural products	Decreased productivity by at least 70% (from 100 cavans per hectare to 15-30 cavans ha-1) High production cost for agricultural products	Decreased productivity by at least 90% (from 100 cavans per hectare to 0-10 cavans/ha-1) High production cost for agricultural products
Floods				
Impacts	Low agricultural production (decreased by at least 40%) and low soil fertility Occurrences of water borne diseases, damages to irrigation infrastructure and siltation	Decreased productivity by at least 40% to 100% Occurrences of water borne diseases, damages to livestock, crops, and properties	Decreased productivity by at least 40% (from 60-90 cavans/ha to 0-30 cavans ha-1) Occurrences of water borne diseases, damages to livestock, crops, and properties	Decreased productivity by at least 50% (from 60-90 cavans per hectare to 0-40 cavans ha-1) Occurrences of water borne diseases, damages to livestock, crops, and properties

4.5. FUTURE IMPACTS AND VULNERABILITY

Streamflow

Relative to the observed data of 1990, the streamflow in PCW is projected to increase all the way through 2080 (Figure 4.4–4.7) under all future climate and land use scenarios. The increase from 1990 ranges from as low as 41.19 mcm to as high as 42.96 mcm per day.

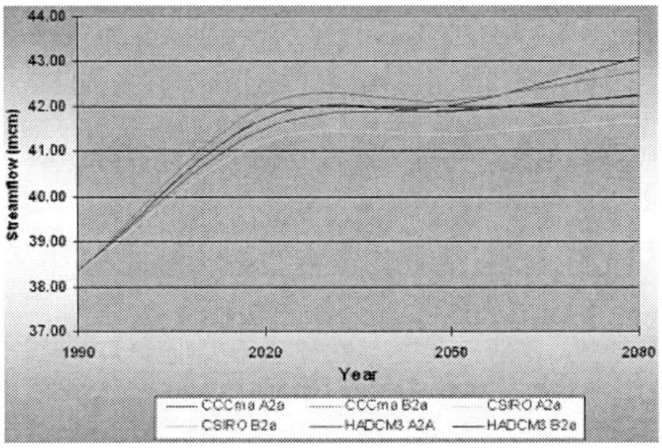

Figure 4.4. Projected average daily streamflow for PCW using 1990 land cover.

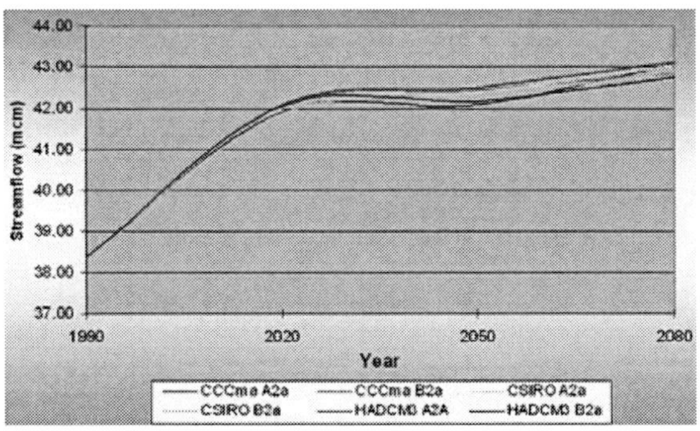

Figure 4.5. Projected average daily streamflow in PCW using 2018 land cover.

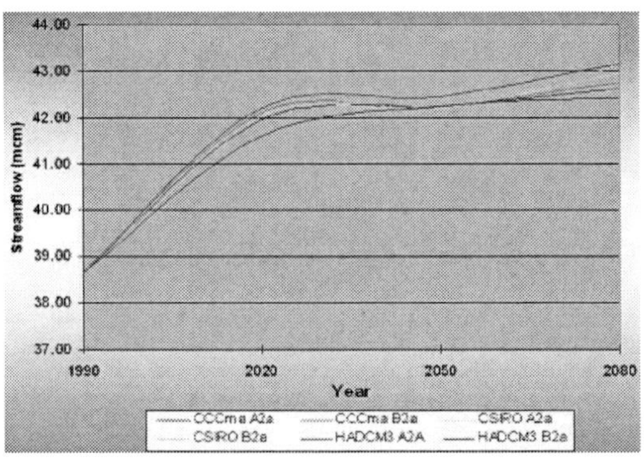

Figure 4.6. Projected average daily streamflow in PCW using 2040 land cover.

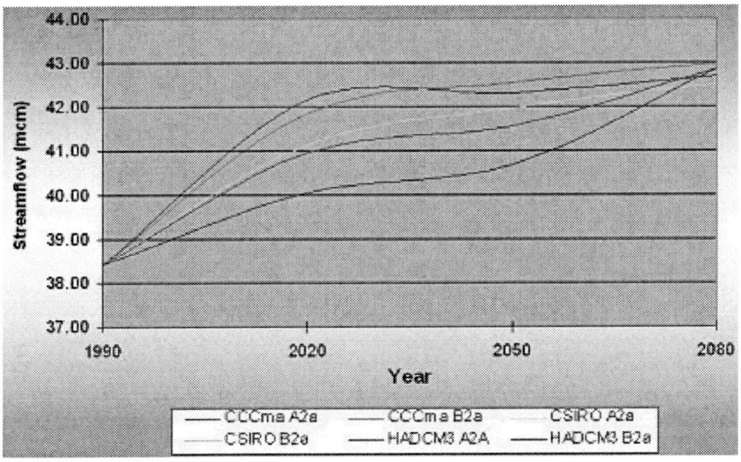

Figure 4.7. Projected average daily streamflow in PCW using 2070 land cover.

For the dry season flow, a declining trend can be noted from 1990 values for all future land use and climate scenarios (Table 4.9, Figure 4.4–4.7). The decreasing streamflow projected for dry season could be attributed to decreasing dry season rainfall and increasing temperature. In effect there will be less water coming into the watersheds from rain but the evaporative demand will increase due to increase in temperature. Dry season flow ranges between 8.84 mcm to 11.2 mcm.

As far as the wet season flow is concerned, there is however, a noticeable increase in the daily average streamflow from 1990 to 2080 (Table 4.10, Figure 4.12–). This could be largely related to the projected increase in rainfall during the wet season. The highest flow is noted 66.45 mcm in 2080 and the lowest projected flow is 63.21 mcm.

The projected alteration in the current streamflow pattern in PCW in 2020, 2050, and 2080 due primarily to projected change in climate, land use and land cover will likely have equally serious, if not graver effects, on the lowland farmers in the service area of UPRIIS compared with the current impacts described above. The projected reduction in the streamflow of PCW during the dry season will tend to heighten the difficulties that farmers are now experiencing. Mainly, this is not because of less water flowing into the reservoir during dry season but more because of the greater need for water and the likely reduction in the water available from the traditional alternative sources such as adjoining rivers and shallow wells.

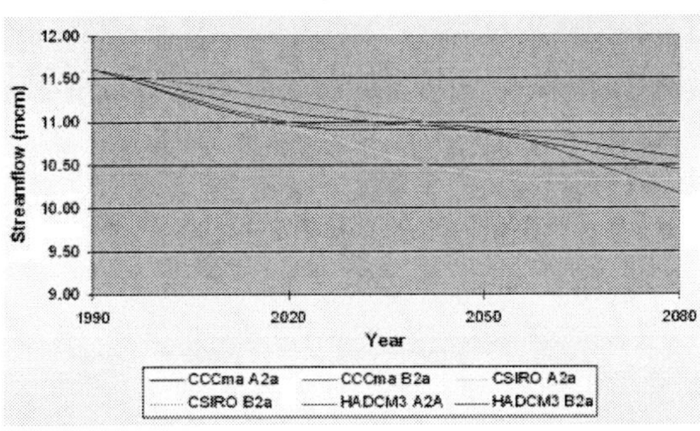

Figure 4.8. Projected daily average dry season flow in PCW using 1990 land cover.

Table 4.9. Projected dry season daily average streamflow (mcm) under future climate and land use scenarios

Year	CCCma A2a	CCCma B2a	CSIRO A2a	CSIRO B2a	HADCM3 A2A	HADCM3 B2a	Average
Land Use 1990							
1990	11.61	11.61	11.61	11.61	11.61	11.61	11.61
2020	11.12	11.25	10.94	11.28	11	10.96	11.09
2050	10.86	10.92	10.37	11.19	10.91	10.89	10.86
2080	10.44	10.87	10.31	11.05	10.15	10.58	10.57
Land Use 2018							
1990	11.36	11.36	11.36	11.36	11.36	11.36	11.36
2020	11.16	11.17	11.05	11.19	11.1	11.08	11.12
2050	10.84	10.73	10.36	10.68	10.75	10.79	10.69
2080	10.53	10.3	10.23	10.21	10.46	10.76	10.41
Land Use 2040							
1990	11.62	11.62	11.62	11.62	11.62	11.62	11.62
2020	11.41	11.34	11.42	11.49	11.35	10.16	11.19
2050	9.98	11.05	10	10.19	10.73	9.63	10.26
2080	9.47	10.76	9.53	9.92	9.86	8.92	9.74
Land Use 2070							
1990	11.4	11.4	11.4	11.4	11.4	11.4	11.4
2020	11.27	10.36	9.24	9.12	10.23	9.68	9.98
2050	10.31	9.63	9.03	7.41	9.99	9.04	9.23
2080	9.6	9.06	8.75	7.29	9.4	8.96	8.84

Table 4.10. Projected wet season daily average streamflow (mcm) under future climate and land use scenarios

Year	CCCma A2a	CCCma B2a	CSIRO A2a	CSIRO B2a	HADCM3 A2A	HADCM3 B2a	Average
Land Use 1990							
1990	57.24	57.24	57.24	57.24	57.24	57.24	57.24
2020	63.32	63.75	62.6	63.93	63.03	63.46	63.35
2050	64	64.13	63.25	64.07	63.79	63.77	63.84
2080	66.12	65.31	63.82	64.81	64.87	64.6	64.92
Land Use 2018							
1990	57.45	57.45	57.45	57.45	57.45	57.45	57.45
2020	63.6	63.79	63.85	63.5	63.94	63.88	63.76
2050	64.12	64.78	64.99	64.52	64.92	64.33	64.61
2080	66.02	65.83	66.21	65.96	66.2	65.31	65.92
Land Use 2040							
1990	57.71	57.71	57.71	57.71	57.71	57.71	57.71
2020	63.48	63.78	63.63	63.78	63.96	63.75	63.73
2050	64.99	64.24	64.9	64.84	64.83	65.26	64.84
2080	66.02	65.33	66.52	66.5	66.64	66.05	66.18
Land Use 2070							
1990	57.49	57.49	57.49	57.49	57.49	57.49	57.49
2020	63.96	64.07	63.65	63.48	62.63	61.45	63.21
2050	64.88	65.8	65.24	65.39	63.81	62.99	64.69
2080	66.05	66.91	66.36	66.19	66.44	66.78	66.45

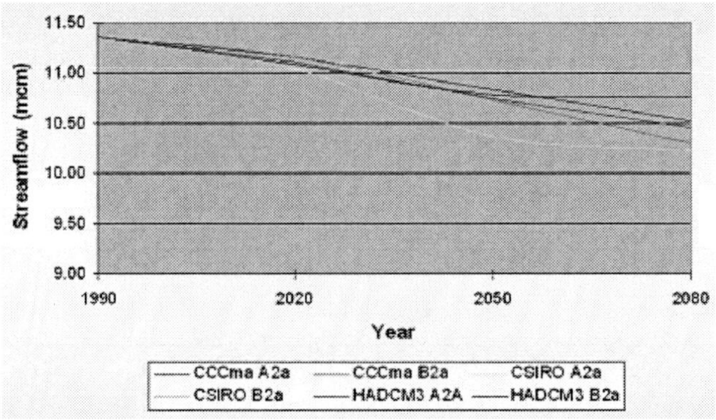

Figure 4.9. Projected daily average dry season flow in PCW using 2018 land cover.

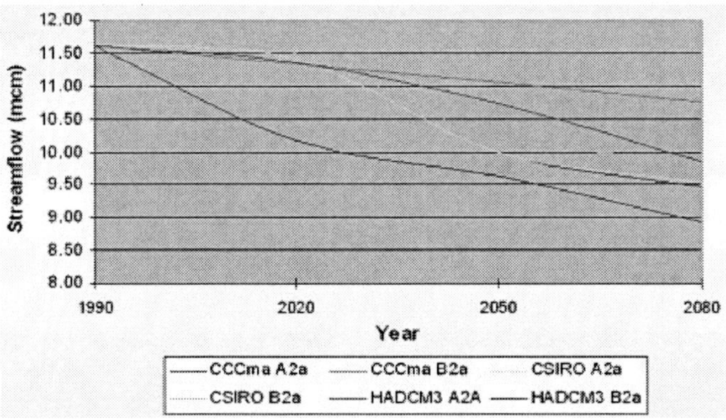

Figure 4.10. Projected daily average dry season flow in PCW using 2040 land cover.

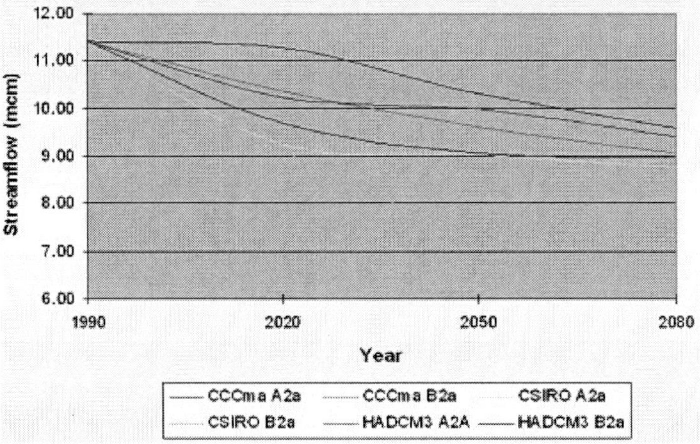

Figure 4.11. Projected daily average dry season flow in PCW using 2070 land cover.

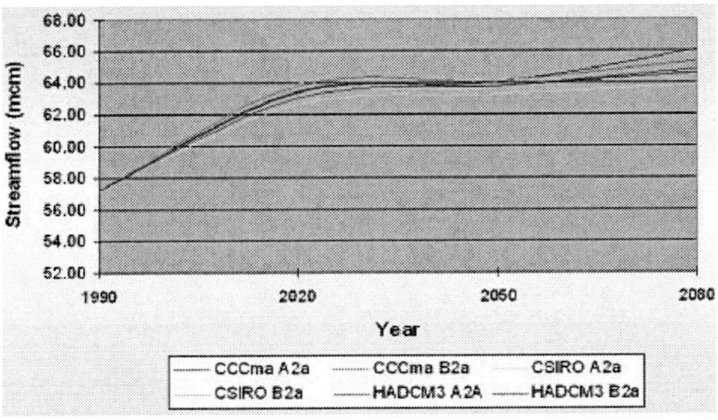

Figure 4.12. Projected daily average wet season flow in PCW using 1990 land cover.

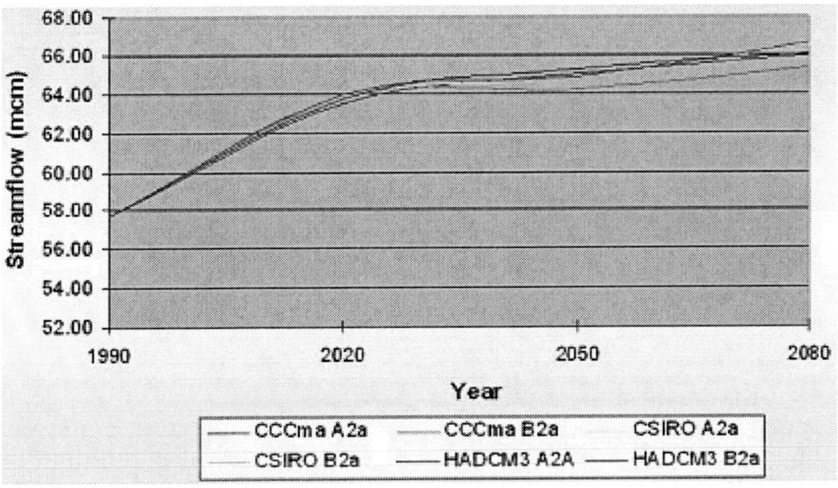

Figure 4.13. Projected daily average wet season flow in PCW using 2018 land cover.

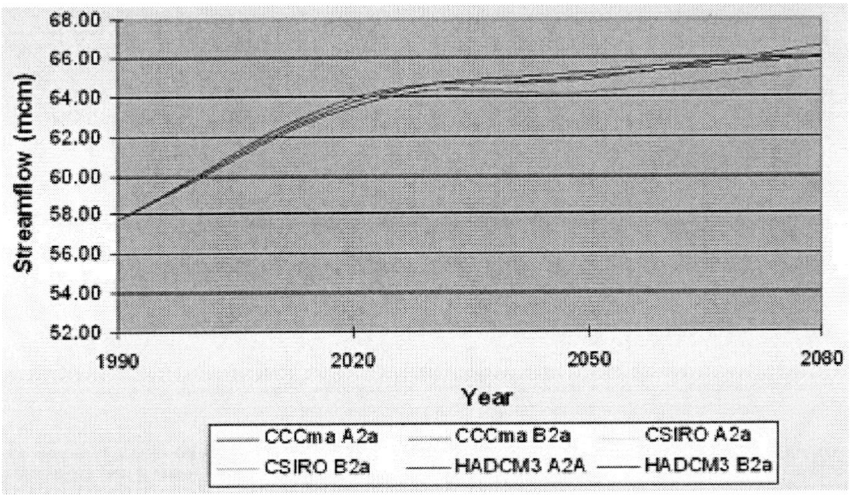

Figure 4.14. Projected daily average wet season flow in PCW using 2040 land cover.

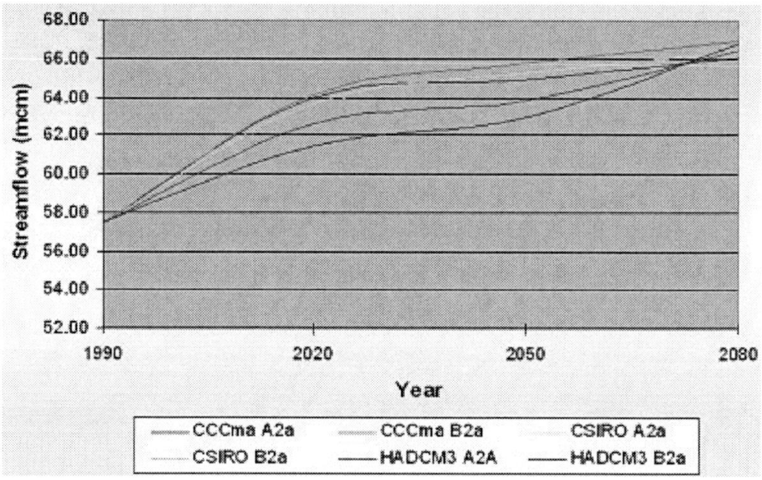

Figure 4.15. Projected daily average wet season flow in PCW using 2070 land cover.

On the other hand, the increasing trend of the wet season flow could likely increase the chances and magnitude of floods in the service area of PCW. Being in the flood plains and inherently vulnerable to floods, most of the service area would be placed at greater risks of being affected by floods that could be magnified by the projected increase in streamflow from PCW. The currently vulnerable sites in each of the service area districts will likely remain to be vulnerable, if not more vulnerable to floods and water shortages, than these areas already are. Most vulnerable will be more than 53,000 ha of the service area that are susceptible to both floods and water shortages (Table 4.11).

Based on the responses of the farmers interviewed during the FGDs and workshops, the basket of options available to farmers who are in areas vulnerable to floods and water shortage is limited to at most three or four options, depending on the location. Given these limited traditional responses to floods and water shortages and the apparent inadequate capacity to fully adopt such measures indicated by the number of farmers who resort to doing nothing and relying on others to meet their needs, and by the absence of financial and other support services from the national and local government agencies, many of the farmers will remain as vulnerable, if not more vulnerable, in the future than they are now.

The projected changes in climate and the associated changes in streamflow patterns of PCW will likely have more serious impacts on the lowland farmers in view of the absence of a deliberate program to reduce the vulnerability of the lowland farmers to floods and water shortages.

4.6. Major Climate Variability and Extremes in PCW

The major climate variability and extremes experienced in the area as identified by the respondents are listed in Table 4.12.

Table 4.11. Vulnerable areas in the different UPRIIS service area districts

District	Service area (ha)	Prone to Flood		Prone to Water Shortage		Prone to Flood and Water Shortage	
		ha	%	ha	%	ha	%
I	24,962.00	7,190.80	28.81	14,580.50	58.41	18,103.30	72.52
II	23,912.90	8,265.30	34.56	8,312.30	34.76	11,325.20	47.36
III	29,846.30	8,489.80	28.45	5,464.90	18.31	11,643.40	39.01
IV	23,811.00	10,223.80	42.94	5,803.00	24.37	12,365.60	51.93
TOTAL	102,532.20	34,169.80	33.33	34,160.70	33.32	53,437.50	52.12

Table 4.12. Major climate variability and extremes identified by respondents from key informant interviews and FGDs

Year	Climate variability and extremes
1974	Typhoon Didang
1978	Destructive typhoon Kading
1979-1980	Drought / El Niño
1982-1983	El Niño
1984	Prolonged rains
1989	Delay on the onset of rainy season
1997-1999	El Niño
2000	Delay on the onset of rainy season
2001	Early onset of rainy season
2002	Delay on the onset of rainy season
2003	Early onset of rainy season

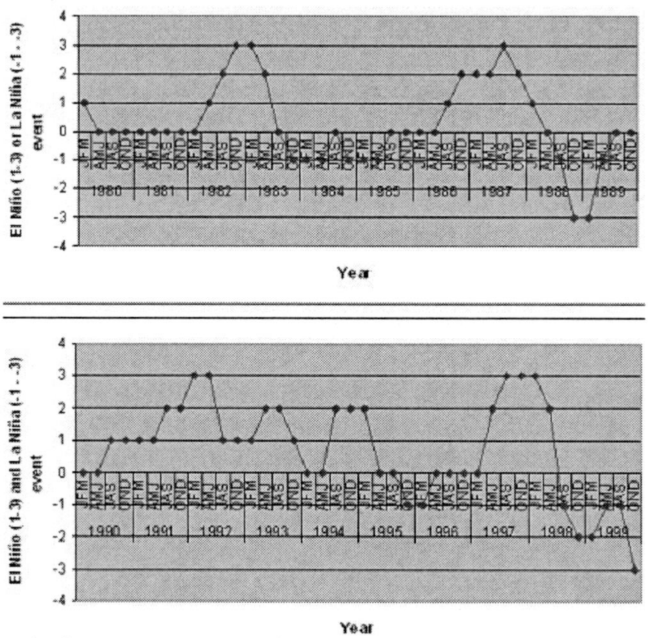

Legend: (3) = strong El Niño event; (2) = moderate El Niño event; (1) = weak El Niño event; (-3) = strong La Niña event; (-2) = moderate La Niña event; (-1) = weak El Niña event; (0) = no El Niño or La Niña event.

Figure 4.16. El Niño and La Niña events recorded by PAGASA from 1980-1999.

The respondents also noted several El Niño episodes, particularly its occurrences in 1979-1980, 1982-1983, and 1997-1999. These observations agreed with El Niño events recorded by PAGASA, as shown in Figure 4.16. Prolonged rains were also observed by the respondents in 1984 which also marked the occurrence of a weak La Niña event (Figure 4.16).

Variability in the onset of rainy season has become a 'common' event since the year 2000. This indicates the unpredictability in the arrival of rains which may be early or delayed. On the other hand, forest fires are frequent in the area and occur yearly since 1980s. Between the years 1980-1988 alone, DENR has recorded an average of 43 forest fires that occurred in PCW yearly, damaging an average of 600 ha of forests a year or a total of 25,783 ha for nine years. Though the high frequency of forest fire incidences coincided with the almost cyclic occurrences of climate variability and extremes such as El Niño and delays in the onset of rainy season, its prevalence cannot be highly attributed to the latter. According to the respondents, most forest fires were intentional since people are practicing kaingin and charcoal making. These practices have become a source of livelihood for them after the termination of the RP-Japan Reforestation Project which provided jobs to the residents.

4.7. Impacts of Climate Variability and Extremes on Local Communities and Institutions

Local Communities

Considering PCW's geographic location, it can be said that all the communities living in PCW are generally vulnerable to climate variability and extremes, including other natural calamities like earthquake. Data available from PAGASA indicate that from 1980 to 1995, a total of 58 strong typhoons, averaging 3.62 typhoons per year, inflicted major damages in the area (Table 4.13). In addition, three major drought episodes occurred during the same period with an average interval of only about three years per episode. These drought episodes occurred in 1983, 1987 and 1991 during which the lowest total annual rainfall and water inflow were registered in the period 1980-2001 (Figure 4.17). This is not to mention the major 1990 earthquake which, besides claiming thousands of lives in Northern Luzon and almost ravaged Baguio into a ghost city, also wreaked havoc on the watershed.

While the exact value of damages inflicted by the past climate-related events in PCW is not available, anecdotal evidences gathered during the survey and FGDs affirm that significant losses have been incurred. These are in the form of claimed human lives, destroyed properties and infrastructures, and damaged sources of livelihood, especially farmlands and fishing areas. Decrease in crop yield is also pronounced in specific years. For instance, records from NIA indicate that in 1990, rice yield fell below average by more than two cavans per hectare in both wet and dry season cropping as a result of drought and typhoons during this year. Local people, however, claim during interviews and FGDs that crop loss could be as much as 100% as a result of droughts and floods. Indeed, some community members are so vulnerable that even before they could fully recover from adverse impacts of previous events, another calamity may strike again that bring them back to the original desperate condition.

Table 4.13. Natural calamities in PCW from 1980-1995

Year	Number of damaging typhoons	Number of drought episodes	Number of earthquake occurrences
1980	2		
1981	3		
1982	4		
1983	3	1	
1984	2		
1985	3		
1986	3		
1987	3	1	
1988	3		
1989	7		
1990	6		1
1991	1	1	
1992	7		
1993	1		
1994	4		
1995	6		

Source: Data from PAGASA.

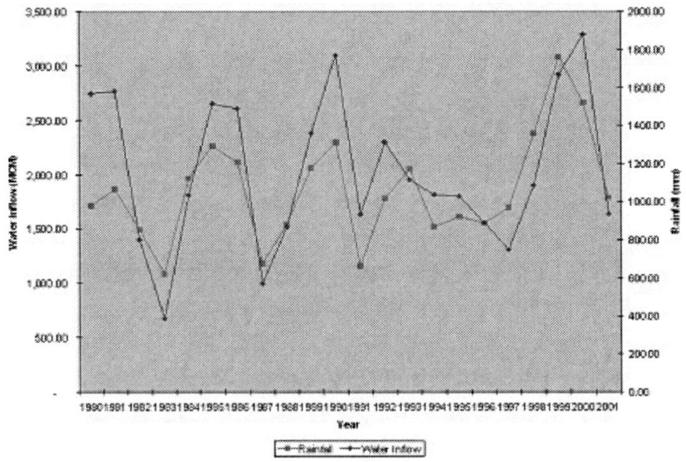

Figure 4.17. Total annual water inflow and total annual rainfall in PCW (1980-2001).

Climate-related events have been observed to have triggered several health problems such as diarrhea, amoebiasis, dehydration, dysentery, dengue, malaria, and typhoid. In the meantime, among the leading causes of morbidity in PCW are respiratory ailments like pneumonia, bronchitis, acute respiratory infection and tuberculosis. Although not yet proven, these diseases may have also been caused by severe climate conditions. Skin disorders are also prevalent in the area which can be attributed to non-potability of water, poor health, and unsanitary practices.

Assessment of vulnerability of the watershed by land use types using the five parameters discussed earlier (slope, elevation, distance from the road, distance from the river, and distance from the community center) with the aid of GIS revealed that more than 65% of the entire PCW is moderately vulnerable to climate variability and extreme while more than 25% is highly vulnerable (Figure 4.18). Most of the areas that are highly vulnerable are forests, grasslands and brushlands by virtue mainly of their location in steep and highly elevated areas and proximity to roads. Areas that are moderately vulnerable are largely grasslands, brushlands, and forests.

On the other hand, among the vulnerable places identified by the local communities themselves during FGDs include low-lying flood-prone settlement areas, agricultural areas prone to floods and droughts, dying streams/rivers, farmlands at the tail-end of irrigation canal, highly erodible areas (in steep slopes) along riverbanks, unstable areas with steep slope that support infrastructure, and grasslands and forested areas/plantations near roads and settlements susceptible to fire.

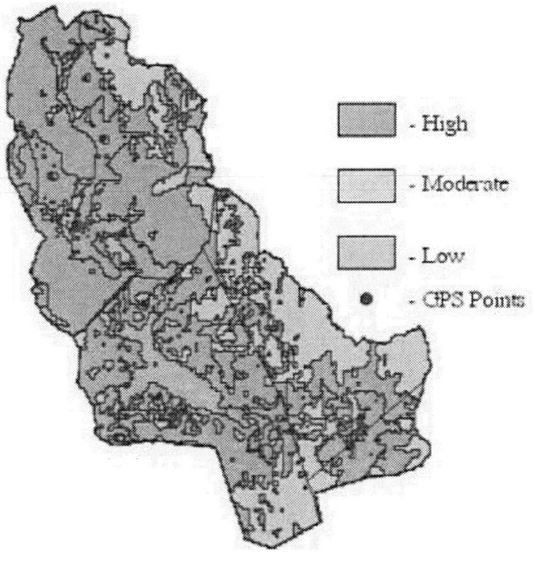

Figure 4.18. Level of vulnerability by land use types and location of vulnerable places as identified by local communities (GPS points).

Plotting the GPS readings of vulnerable places identified by the local communities during FGDs in the vulnerability map generated through GIS provides an interesting result.

Table 4.14. Number of GPS readings of vulnerable places identified by the local communities per municipality in PCW that fell within the different vulnerability levels generated through GIS

Province / Municipality-(No. of GPS Readings)	Vulnerability Level			Total GPS Readings
	Low	Moderate	High	
Nueva Ecija				
Carranglan	6	29	0	35
Pantabangan	2	21	2	25
Nueva Vizcaya				
Alfonso Castañeda	0	8	9	17
Aurora				
Maria Aurora	1	6	2	9
Total GPS Readings	9	64	13	86

As shown in Table 4.14, there is high congruence between the GIS-generated levels of vulnerability with the vulnerable places identified by the local communities. Sixty-four of the 86 GPS readings or about 74% fell within the moderate vulnerable areas while 15% and 11% fell in the high and low level categories, respectively. This implies that the approach of combining the two methods of identifying vulnerable areas could be a useful tool to provide a more comprehensive assessment of vulnerable areas/places.

Extent and Nature of People's Vulnerability

Based on the vulnerability index developed from the results of the household survey, farmers in general are more vulnerable to climate variability and extremes compared to non-farmers (Table 4.15). This is regardless of the source of weights used in the index, i.e., whether determined by the researchers or the local communities themselves. However, the index developed using the researchers' weights produced both the highest (66.53) and lowest value (4.37) compared to the weights provided by the local communities (59.12 for maximum and 11.8 for minimum value).

Indeed, the values of the index are relative since they are sensitive to the perceptions of whoever is giving the weights. There is therefore the need to involve the different stakeholders in coming up with vulnerability index, particularly those directly affected by climate variability and extremes like the local communities, to ensure the appropriateness of the index.

Table 4.15. Values of vulnerability index for farmers and non-farmers based on the weights provided by researchers and local communities

Source of Index's Weights	No. of Respondents	Vulnerability Index (Possible Value from 0-100)		
		Mean	Minimum	Maximum
Researchers				
Farmers	70	38.14	6.87	66.53
Non- Farmers	38	24.56	4.37	43
Combined	108	33.37	4.37	66.53
Local Communities				
Farmers	70	42.87	18.95	59.12
Non- Farmers	38	26.3	11.8	55.12
Combined	108	37.04	11.8	59.12

On the other hand, local communities are themselves very heterogeneous. During the FGDs, the local community members identified at least three categories of farmers as well as other socioeconomic groupings in PCW that have varying degrees of vulnerability to climate variability and extremes. These are 'small', 'average', and 'rich' farmers, fishermen, employees, and small business entrepreneurs. The group showing evidences of being the most vulnerable based from the FGDs and households surveys are the small farmers. Most of them have very low educational attainment, do not own a parcel of land, have very meager income, without capital, and do not have access to other productive resources. Some may even live in vulnerable places and have ineffective adaptation strategies to variable and extreme climate conditions. This group is considered the most vulnerable because even if the overall climate-related losses may not be that devastating at the community level, the damage it creates to the household could have lasting impacts and could lead to a chain-reaction of negative effects.

The group considered to have moderate vulnerability is composed of fishermen, farmers with small land and little capital, owners of small enterprises, sawali makers, and employees of various agencies. They are better educated compared to the small farmers. They may also have access to productive resources such as land, capital, and technology but do not have control over them. Despite this, however, few of them may have income below the annual per capita poverty threshold recorded at 13,843 pesos for the Central Luzon Region (NSCB 2000). Some from this group may also live in vulnerable areas such as in low-lying flood-prone places and those where sources of water are limited in case of drought. Compared to the most vulnerable group, they are relatively less sensitive to climate-related losses due to their access to limited resources and relatively better adaptation strategies.

The last group considered to be the least vulnerable constitutes the rich farmers and the overseas workers. Affluent farmers in general are the most educated among the three groups.

Table 4.16. Vulnerability of various socioeconomic groups to climate variability and extreme based on certain characteristics

Description	Socioeconomic Groups			
	Small farmers	Average farmers & fishermen	Employees/ Small entrepreneurs	Rich farmers
General socioeconomic characteristics	Mostly with low educational attainment, no farm land and capital, very low income, almost no access to other productive resources	Finished elementary education or reached high school, some access to productive resources such as land, capital, and technology	College or high-school graduates, some access to productive resources such as land, capital, and technology	College or high-school graduates; more access and control over productive resources including appropriate linkages
Nature of impacts of climate variability and extremes	Decline in crop production, food, livelihood, health condition; more debt incurred	Decline in crop/fish harvest and income, food availability, livelihood sources; health condition may or may not be affected	Increase in prices of commodities; limited sales	Decline in production and income; no change in food availability, livelihood and health
Degree of negative impacts	High	Moderate	Moderate	Low
Examples of adaptation strategies	Avail of high interest loan or borrow from relatives; plant vegetables along rivers/plant other crops; work in nearby towns; engage in other jobs	Plant vegetables along rivers/plant other crops; engage in other sources of livelihood	Avail of government loans; engage in backyard project; store food supply and other farm inputs for sale	Store food and farm inputs
Effectiveness of adaptation strategies	Some effective, others not	Effective	Effective	Effective
Location of settlement/properties relevant to vulnerable areas	Some are located in vulnerable areas	Some are located in vulnerable areas	Some are located in vulnerable areas	Generally located in secured areas
Degree of vulnerability	High	Moderate	Moderate	Low

Table 4.17. Vulnerability of various socioeconomic groups to future climate variability and extremes based on certain characteristics

Description	Socioeconomic Groups			
	Small Farmers	Average Farmers and Fishermen	Employees/Small Entrepreneurs	Rich Farmers
Food availability and crop production	Decline in crop production; starvation	Decline in crop production and other livelihood resources	Increase in prices of commodities, hence increase in expenditures	Supply of food is not affected because they have money to buy
Livelihood	Worsening poverty condition; more debts incurred and longer time to repay	The livelihood sources of some will decline, while others will improve, especially those who loan money to the poor farmers with collateral.	Decline in business activities of small entrepreneurs, and limited money to spend due to increase in prices of commodities. However, they are not much affected because some have alternative sources of livelihood, like livestock raising.	They become richer because they obtain the farms and other possessions (collateral) of the poor who loaned money and was not able to repay. The poor farmers also approach them for farm inputs which they return with interest. Rich farmers are also the buyers of "palay", hence they have control over crop prices.
Health	Their health will be affected by intense climate condition and malnutrition. Since they don't have money to consult a doctor or buy medicine, they will just resort to medicinal herbs or consult an "albularyo".	Their health will not be much affected.	Their health will not be much affected.	Their health will not be much affected.
Water supply	Shortage in water supply for farm and domestic uses. The assistance given by government in terms of water pump usually does not reach them.	Some will experience water shortage while others will not be much affected because they have money to buy water for domestic and drinking purposes, as well as water storage facilities.	Expenditures for water will increase, but their water supply will not be much affected because they have money to buy water for domestic and drinking purposes, as well as water storage facilities.	Water supply will not be much affected because aside from the capacity to make/find alternative sources of water, they also have money to buy water for domestic and drinking purposes, as well as water storage facilities.

Table 4.17 (Continued)

Degree of negative impacts	High	Moderate	Moderate	Low
Examples of adaptation strategies	They work in other farms, engage in other jobs, work in nearby towns, or even apply for jobs abroad. They also plant crops that can adapt to the dry season, like onions and tomatoes. Others make "sawali" from cogon grasses that can be harvested in the mountain.	They plant fast growing crops and store food supplies. They also invest in other businesses or find other sources of income. They look for job in other places.	They decrease budget in some expenditures and store food supplies.	They will be selective of whom to lend money to. They plant crops in other areas where there is water. They store food supplies.
Effectiveness of adaptation strategies	Some effective, others not	Some effective, others not	Some effective, others not	Some effective, others not
Location of settlement/ properties relevant to vulnerable areas	They have no choice but to stay in their area because they don't have the capacity to transfer to safer locations. In times of extreme weather events like typhoons, they need to evacuate to safer areas like schools.	They have the capacity to select or transfer to safer locations. Also, most of them live in high and safe places and their homes are made of sturdy materials like concrete.	They have the capacity to select or transfer to safer locations. Also, many of them live in safer places and their houses are made of sturdy materials like concrete.	They have the capacity to select or transfer to safer locations. Their houses are located in safer places and are made of concrete. There are some who have houses in other places.
Degree of vulnerability	High	Moderate	Moderate	Low
Present distribution of farmers	75-85%	5-15%	5-10%	2-5%
Probable distribution of farmers in the future	85-90%	5%	5-8%	2-5%

They usually own large tract of land/farm, possess investment capital, own farming machineries and tools, and have control over other factors of production including technology. They also live in favorable areas which are less susceptible to flooding and have effective adaptation strategies.

On the other hand, overseas workers are also better educated like the more affluent farmers. They have some access to financial resources and have linkages with other institutions outside the community. Their families are considered among the least vulnerable group because the financial support they provide is fixed and stable and unaffected at all by variable and extreme climate events in the local area. Similar to the well-off farmers, their families also live in safer places and have better adaptation strategies.

Table 4.17 presents a detailed description of the different socioeconomic groups in PCW relating to their degree of vulnerability to climate variability and extremes. It should be highlighted that among the three groups, the small farmers in general may in fact have the most number of adaptation strategies. Some of these strategies, however, like availing high-interest loan, are ineffective thereby only increasing their degree of vulnerability.

4.8. Vulnerability of People to Future Climate Variability and Extreme

Various climate change scenario models have predicted decrease in precipitation and increase temperature in PCW in the next 20, 50, and 80 years. This is expected to further cause negative effects in food availability, crop production, livelihood, health, and water supply of the residents in the watershed.

Further increase in temperature and decrease in precipitation presents a gloomy scenario to the small farmers in PCW. Not only will their crop production decline, but starvation is likely to be experienced which would result to malnutrition and other kinds of diseases. Many of them will be engaging in other jobs since the farms that they used to tend will be confiscated because of unpaid debts. Availing of high interest loans may no longer be an available adaptation option since they do not have collateral to guarantee for loans. Moreover, lenders will also be selective of their clients and will provide loans only to individuals who have the capacity to pay. The poor farmers also have no choice but to stay in their area because they do not have money to transfer to other locations. Hence, in times of extreme weather conditions like typhoons, they need to evacuate to safer places like schools.

Meanwhile, average farmers, fishermen, employees, or small entrepreneurs will still have moderate vulnerability to probable increased temperature and decreased precipitation in the future. Though their food, livelihood, health, and water supply may be affected by future climate variability and extreme, a few adjustments on their expenditures and other activities will enable them to cope with the negative impacts. Should the need arises, they also have the capacity to transfer to less vulnerable places in times of extreme weather conditions.

Finally, the rich farmers, although slightly affected by the probable changes in climate, appear to benefit more from the situation. This will be in terms of increased farmlands and other possessions from the collateral of the poor farmers who will not be able to repay their debts.

As implied in the previous sections, while PCW communities are generally vulnerable to climate variability and extremes by virtue of their geographic location, their degree of vulnerability varies based on a combination of other factors. These factors include the farmer's or household's socioeconomic circumstances as well as the broader sociopolitical and institutional contexts.

4.8.1. Socioeconomic Factors

Table 4.18 presents the significant factors associated with vulnerability based on Spearman correlation analysis considering the weights provided by the researchers and the local communities. Using the researchers' weights, three factors have significant correlation with vulnerability: farm income, monthly food expenditures, and farm distance to market. In the case of the first factor, people with high farm income have the tendency to be more vulnerable compared to those with low farm income. This implies that the more dependent people are in their income from the farm, the more they become vulnerable to climate-related disorders. On the other hand, the variable on monthly food expenditures is negatively correlated with vulnerability. This means that people who spend less on food, presumably because they have limited financial resources, are likely to be more vulnerable to adverse climate conditions. Finally, farm distance to market is also positively correlated with vulnerability, although the degree of association is quite weak (at 0.05 confidence interval). As learned during the field work, households from far flung areas are cut off from markets during rainy season and flooding that make them more vulnerable.

Similarly, three factors are also found to have significant relationship with compliance using the weights provided by the local communities themselves, namely: number of organizations joined, farm size, and monthly food expenditures. A positive relationship exists between the number of organizations joined by the farmers and their vulnerability. This implies that what really matters in terms of reducing vulnerability is not the quantity of organizations joined by the farmers but the quality of services provided by these organizations. Similarly, farm size is positively correlated with vulnerability, meaning the larger the farm size owned by the household the more they are likely to be more vulnerable. This can be explained by the fact that most farmers in PCW usually devote their farms into a single commodity, rice, thereby making them more vulnerable to variable and extreme climate conditions.

To identify and evaluate the combination of factors that significantly affect the households' vulnerability, vulnerability index was regressed with the different predictor variables. Out of the 17 postulated predictor variables, five variables were found to be significantly related with households' vulnerability using the weights provided by the researchers (Table 4.19). These were: sex and ethnic affiliation for demographic factors, number of organizations joined and land ownership for socioeconomic factors, and farm distance to market for geographic factor.

Table 4.18. Correlation coefficients between the postulated factors and vulnerability

Postulated factors	Weights by Researchers		Weights by PCW ommunities	
	Vulnerability coefficients	Level of significance	Vulnerability coefficients	Level of significance
1. Demographic				
age	-0.07935		-0.1208	
gender				
ethnic affiliation				
educational attainment	-0.06391		-0.0398	
household size	0.01438		0.0015	
2. Socioeconomic				
total income	0.03483		0.0266	
household asset	-0.1782		-0.0845	
number of organizations joined	0.18399		0.2205	0.05
farm size	-0.1199		0.3241	0.01
farm income	0.26165	0.01	0.4393	
number of transportation system	-0.07328		-0.0168	
monthly food expenditures	-0.29576	0.01	-0.295	0.01
no. of loan applied	0.06742		0.12755	
no. of information sources	0.01012		0.1116	
3. Geographic				
farm distance to market	0.24182	0.05	0.212	
4. Overall coping mechanisms				
number of coping mechanisms	-0.08644		0.0282	

In terms of demographic factors, women are found to be more vulnerable compared to men while migrants are more vulnerable than native inhabitants. The vulnerability of women may be attributed not only to their limited physical capacity but also since they have to bear most of the family burdens associated with climate variability and extremes like borrowing money and/or making both ends meet in case of crop failure and caring for the sick children. These, they have to do, on top of the already burdensome household chores that they have to religiously perform. On the other hand, the migrants' vulnerability may be related to their unfamiliarity to the area, hence they are unable to better prepare or develop appropriate

Table 4.19. Coefficients of the postulated predictors of household's vulnerability by step-wise regression analysis

Postulated Predictors	Code	Weights by Researchers		Weights by PCW Communities	
		Regression Coefficient	Level of Significance	Regression Coefficient	Level of significance
1. Demographic					
age	AGE				
gender	SEX	-9.66	0.01		
ethnic affiliation	NATIVE	-10.11	0.01	-0.2907	0.01
educational attainment	EDUC				
household size	HHMMDEPD			0.2781	0.05
2. Socioeconomic					
total income	TOTNCOM				
household asset	HHASSET				
number of organizations joined	NORGJ	9.74	0.01		
farm size	FARMSIZE				
farm income	FARMNCOM				
number of transportation system	NTRANSP				
monthly food consumption	FUDMON			-0.3929	0.01
no. of loan applied	NLOAN				
no. of information sources	NFOSURZ				
land ownership	LANDOWN	-8.3	0.05		
3. Geographic					
farm distance to market	FRMDSTMK	0.0006	0.01	0.401	0.01
4. Overall coping mechanisms					
number of coping mechanisms	NOCOPING				
Intercept		46.25		43.73	
Coefficient of determination		0.46		0.43	

Note: Variables without corresponding coefficient values do not meet the 0.05 level of significance for entry into the model.

In terms of geographic consideration, farm distance to market is positively and significantly related to vulnerability. This affirms the significant relationship between these two variables using the correlation analysis.

Using the weights provided by the communities, four variables are found to be significantly related with households' vulnerability: ethnic affiliation, household size, monthly food consumption, and farm distance to market. Two of these variables, namely, ethnic affiliation and farm distance to market, were also found to be significantly related with the household's vulnerability using the weights provided by the researchers. On the other hand, bigger-sized households are likely to be more vulnerable compared to smaller-size households probably because the former have more mouths to feed compared to the latter. Moreover, monthly food consumption is found to be negatively and significantly related with vulnerability. This affirms the output of the correlation analysis that households who are unable to spend much for food potentially because they have limited financial resources are inclined to be more vulnerable than those who can spend more.

Based on the computed coefficient of determination, about 46.25% and 43.73% of the total variation in vulnerability rating using the weights provided by the researchers and the local communities, respectively, are explained by the above mentioned significant variables (Table 4.19). This means that an average of around 55% of the vulnerability variance based on the weights provided by the two groups are still unaccounted for on an aggregate level. There is, thus, the need to look for other factors that may help explain the households' vulnerability aside from those identified in the regression model.

4.8.2. Contextual Factors

In addition to the above mentioned significant factors, the broader sociopolitical context by which the communities participate influence their level of vulnerability. As mentioned earlier, the chain of development projects implemented in the area from 1971 to the present have in some ways created a sense of dependency in the part of the local communities for external assistance. This is because these projects, particularly the resettlement one, were more of a dole-out in their orientation with very little attempt towards building local capacities. Consequently, the culture of self-reliance was not fully developed contributing to the vulnerability of some members of the local community especially with the termination of these projects.

Similarly, there is lack of enabling national policies and institutional support that could help reduce the local communities' vulnerability and enhance their adaptive capacity to minimize the adverse impacts of climate variability and extremes. For instance, the government forest policy does not allow timber harvesting in all watershed areas that support big infrastructure projects such as PCW even if the communities themselves are involved in plantation establishment. This has discouraged their active participation in reforestation and forest protection activities and has led, in many cases, to deliberate burning of established plantations. In the absence of direct benefits from established plantations and due to limited sources of livelihood opportunities in the area, community members are compelled to engage in illegal cutting and charcoal-making to augment their meager income that has led to the degradation of some parts of the watershed, contributing to its biophysical vulnerability. Similarly, despite the presence of the different institutions in the area such as NIA, NPC, and

DENR, whose main focus is to protect their investments, the interest of the local communities is seen only as secondary priority. Through time, there has been declining support towards community development and the provision of more sustainable sources of livelihood. Moreover, institutional support to anticipate and adequately plan for the occurrence of variable and extreme climate conditions is yet to be developed. Similarly, there are yet to be initiatives directed towards enhancing current adaptation strategies and capacity building at the local level.

Finally, the prevailing inequity that characterizes the Philippine social structure is very much evident in PCW that further contributes to the vulnerability of the poor community members. The community's own typology of small, average and rich farmers is a concrete reflection of the inequitable social structure that prevails in the area. As already mentioned the well-off sector of the community has better access and control over productive resources and has the option to live in safer places putting them at a less vulnerable situation. The same sector is also more inclined to capture most of the benefits from the different development projects due to better association and linkage with institutions that implement these projects.

4.8.3. Institutions

Table 4.20 presents the positive and negative impacts that climate variability and extremes bring to the institutions. One positive impact mentioned especially during the occurrence of rains is the increase in the level of water in the reservoir which is good in power generation and in irrigation. Consequently, after the rains soil conditions and other elements become conducive to planting. The LGU also mentions that in times of typhoon, they are given a share of 5% from the Internal Revenue Allocation (IRA), which they use to minimize the negative impacts of the events.

Meanwhile, the negative impacts of climate variability are discussed below:

Operation

Most respondents indicated that the presence of climate variability and extremes hamper their operation. This is due to the lack of sufficient water in the reservoir (especially in times of El Niño), the necessary resource for NIA and NPC's operations. Meanwhile, DENR mentions that the presence of El Niño leads to the decline of their tree planting activities and it increases the occurrence of forest fires. LGUs, on the other hand, are concerned with finding solutions to help the communities have sufficient supply of water for their daily activities. Most of the interviewees from the Barangay Development Council (BDC) responded that El Niño has no significant effect on them because they are already adjusted to many climate variability and extremes they have been experiencing. Only three out of 20 (15%) respondents from BDC said that the presence of climate variability and extremes affected them (Gen Luna, Lublub, and East Poblacion.).

Table 4.20. Institutional impacts of climate variability and extremes

Nature of CV&E / Stakeholders	Operations	Budget	Manpower	Programs	Others
Typhoons					
NIA Positive effects :full resource reservoir reached the maximum level	Cannot deliver water due to the destruction of canals	Decrease in tax collection due to non-payment of fees for irrigation water	None	Delayed implementation of programs for rehabilitation	Cannot deliver irrigation services to other areas Destroyed infrastructure
DENR Positive Effects: After the typhoon, the condition is conducive for planting and it lessened the occurrence of forest fire	Destruction of forest plantation roads Difficulty in field work and supervision	Needs additional budget for forest rehabilitation	Addition of casual employee	Intensive rehabilitation programs were formulated and implemented sometimes leads to terminating jobs/ programs	Difficulty in transporting seedlings from the top (mountains)
LGU	Increased in operation cost	Lack of budget	Needs additional manpower	Intensive rehabilitation, socioeconomic and infrastructure programs	Additional services and cost
NPC Positive effects:Increase water in the reservoir	Transmission lines were destroyed w/c hamper power generation	Higher restoration cost of emergency lines	Additional hiring	Give way to most urgent programs and needs	Decrease in power generation
Barangay Development Council (BDC)	Cannot travel to the municipal hall to ask help due to floods (Gen. Luna)	Insufficient budget (Lublub)	None	None	None
Drought/ El Niño					
NIA	Decline in level of water in the reservation	Shortage of budget due to non-payment of fees for irrigation water	Delay in salary/wages of casual employee)	Decrease in program target Production in water supplied to service area decreased area covered by the regular operation	None

Table 4.20 (Continued)

DENR	Increase in the occurrences of forest fires Delayed planting of trees	Additional budget requirements to minimize slash and fire	Additional manpower	Focus on forest protection	Low survival rate of established nursery/plantation
LGU	Additional projects or activities, e.g., establishment of water impounding projects (irrigation) Searching for potential spring development (SPSP)	Additional budget/funds needed for the spring development	None	Some programs activities either delayed or set aside to give way to additional water related projects	Some farmers are not able to pay (tax) resulting to a reduction in water supplied to the service area
NPC	Decline in available water in the reservoir resulting to decline in power generation	Special addition of budget to offset financial request	None	Delayed activities (e.g., reforestation) Adjustment of program schedules	Shortage in electricity delivered
BDC	None	Increased expenses (Lublub & Cadaclan) Insufficient funds (East Pob.)	None	Destroyed water pump (Fatima) Delayed project implementation (East Pob.)	None
NIA	Increase area of responsibility in the forest/ plantation	Additional budget required	Additional manpower	Change in priorities Additional areas to be replanted supply gaps	Diversion of focus of normal/ irregular activities to control forest fires
Forest Fires					
DENR/NPC and LGU-BDC	Reduction in the refo project accomplishment	Additional budge Increased in labor cost	Additional manpower Mobilization of forest guards & fire brigades Training of people	None	None
Delay onset of rains					
DENR	Disruption of planned schedules				
NPC	Shortage of available power			Delayed training activities	
LGU-BDC		Decreased in tax collection (Conversion) Additional funds (Galintuja, Ma. Aurora)			

Early onset of rains					
DENR	Decreased survival rate in plantation	None	Deplete manpower	None	None
NPC		(+) effects triggers more power generation		Delayed construction of structures	
BDC/ LGU	None	None			
NIA		None			

The presence of typhoons also bring negative effects to the institutions. Some of these effects are the destruction of irrigation canals that lead to the problems in water delivery of NIA, destruction of forest plantation roads resulting to delays and difficulties in the reforestation activities of DENR, and fallen transmissions lines of NPC due to strong winds and rains that affect supply of power to the communities. LGUs also say that typhoons increase the cost of their operation due to additional manpower/workforce, equipment, and instruments needed to help the communities.

On the other hand, forest fires result to a decrease in the accomplishment of many institutions in terms of the number of trees planted (DENR, NPC, LGU, and BDC). Respondents from NIA also mention that this event increases their area of responsibility because the fire affects not only the area where the fire occurs but also the whole watershed. Delay on the onset of rainy season also affects the institutions negatively. In the case of DENR, it disrupts the planned schedules of tree planting. Many do the planting before the rainy season because newly planted trees need a lot of water for their growth. However, when the rainy season starts early this results to a decrease in the survival rate for the established plantations.

Budget

Many respondents emphasized that the presence of climate variability and extremes resulted to an increase in their expenses; hence, they need additional budget. The additional amount will be used to repair properties destroyed, procure new equipments, and continue their operation.

Manpower/Workforce

Almost all the institutional respondents mention that the occurrence of climate variability and extremes such as typhoons and El Niño results to decreased manpower/workforce and triggers the need for additional hiring. Insufficient manpower/workforce affects the institutions in carrying out activities efficiently and effectively, especially in solving the problems they encounter. They need additional labor to continue their operations and execute and implement planned programs and schedules. However, some institutions like DENR experience difficulty in hiring casual laborers if rainy season starts early. This is because most of the people in the area are farmers who have to attend to their farms to ensure that it will have enough water.

Programs

The different programs of the institutions, both planned and implemented, are also affected by the presence of climate variability and extremes. These events cause some delays and changes in the programs. Sometimes, institutions have to terminate or totally cease the operation of a program to give way to the most urgent problems needing urgent solutions. Such problems include reforestation, tree planting, small water impounding projects/programs (SWIP) during the times of El Niño and forest fires, and rehabilitation activities (infrastructures, etc.) for typhoons.

Box 2. Summary for decision makers

Climate change may have positive or negative impacts to different forest ecosystems in PCW. The degree of impacts of climate change varies depending on the type and location of forest ecosystems being affected. La Niña and El Niño have the most negative impacts to forest ecosystems. The lowland farms, grasslands, and natural forests are the most vulnerable among the forest ecosystems.

- The small or poor farmers are the most vulnerable group in PCW. These are farmers who do not own land, have no capital, are without farming machineries, live in vulnerable areas, and whose adaptation strategies are ineffective.
- In indexing, assigning weights affects outcomes.
- Looking on the multiple stressors both at the micro and macro levels (scale issue) that contribute to people's vulnerability is a useful way of understanding this complex concept.
- There is a need for bottom-up assessment and planning to address vulnerability and enhance adaptive livelihood at the local level. Participatory action research engaging the different stakeholders should be pursued to minimize vulnerability of the poor and enhance adaptive capacity at the local level.

To reduce vulnerability, policies and development programs should aim at empowering the local communities to broaden their range of choices of appropriate strategies rather than making them dependent on external support. These policies and development programs should not preclude questioning the large scale structural causes of vulnerability such as poverty, inequity, institutional, and economic barriers to development. Also, it should not neglect the issue of power and conflict (Brooks 2003).

The impacts of the changing streamflow patterns in PCW on the lowland rice farmers served by the UPRIIS were characterized largely through FGDs, key informant interviews and workshops that heavily involved farmers, key personnel from NIA, NPC, LGUs and local community leaders, among many stakeholder groups in PCW. Review of relevant documentation was also useful in this part of the study.

The increase in streamflow could lead to higher likelihood of floods in the service areas of UPRIIS than it is at present. Likewise, the projected decrease in streamflow of PCW during the dry season will likely increase the incidence of water shortage which could be aggravated by the increasing water demand due to increasing temperature.

Others

Many of the respondents share that the delivery of the services coming from them is also affected by climate variability and extremes. It leads to difficulty in delivering water for irrigation (NIA) and decrease in power generation (NPC), resulting to occurrence of blackouts in communities. It also causes difficulty in transporting seedlings from the nursery to the reforestation sites, lowering of survival rates of the established nurseries (DENR), and reduction of local government IRA collections because of the burden experienced by the communities.

It can be noticed that the different impacts of climate variability and extremes to the institutions exhibit a ripple effect. When the operation of institutions is affected negatively, budget allocation, manpower/workforce, program implementation, among others, are also negatively affected.

It is also noteworthy that either the delay or onset of rainy season is not given much attention by the institutions. Instead, what concerns them a lot are the presence of El Niño or prolonged rains because these cause more damage or problems in their operation, manpower, budget, and programs.

Chapter 5

Adapting to Climate Change

The preceding chapter highlights the potential impacts of climate change to forest ecosystems, water resources and local communities. This chapter discusses potential solutions that address these impacts by adapting to a changing climate.

Climate change adaptation refers to '"adjustment in natural or human systems in response to actual or expected climatic stimuli or their effects, which moderates harm or exploits beneficial opportunities"' (IPCC 2001). On the other hand adaptive capacity is the property of a system to adjust its characteristics or behaviour, in order to expand its coping range under existing climate variability, or future climate conditions (Brooks and Adger 2004). In practical terms, adaptive capacity is the ability to design and implement effective adaptation strategies, or to react to evolving hazards and stresses so as to reduce the likelihood of the occurrence and/or the magnitude of harmful outcomes resulting from climate-related hazards.

Climate change adaptation does not occur in a vacuum. Ideally, adaptation strategies should be part of the sustainable development agenda. One reason why climate change is neglected is that there are more pressing concerns such as poverty, health issues, economic development, and environmental pollution. However, a strong case can be made that adapting to climate change leads to sustainable development. In other words, failure to adapt to a changing climate will ultimately lead to unsustainable development whether at the national or local level. For example, by increasing resilience of local communities to cope with climate risks, they are also better prepared to face the different climate in the future.

Adapting to climate change is best pursued by building on coping mechanisms to current climate variability and extremes. The Philippines has had a long experience in coping with climate risks such as tropical storms and ENSO phenomenon. Adaptation strategies for these climate risks can be used as building blocks for long-term climate change. For example, farmers have evolved adaptation strategies to strong rains caused by the La Niña phenomenon. Since climate models predict higher rainfall intensity for many parts of the Philippines, adaptation strategies for La Niña could be useful in developing adaptation to higher rainfall brought about by a warming climate.

5.1. Forest Ecosystems

5.1.1. National Policy Framework and Potential Adaptation Strategies

To date, there has been little consideration of an overall climate change adaptation strategy and its various options for Philippine forest ecosystems. The 1999 PINC contains adaptation options for watershed management that partly apply to forest ecosystems. These are mainly contained in the laws and polices governing the use and conservation of forest resources in the Philippines. The more pertinent of these include the following:

- Presidential Decree 705 of 1975 (Revised Forestry Code of the Philippines) embodies the general mandate of the Constitution in managing and conserving forest resources.
- DENR Administrative Order No. 24 Series of 1991 promulgates the shift of logging from old-growth forests to secondary (residual) forests effective 1992. Prior to this, logging was confined to old-growth forests.
- Republic Act (RA) No. 7586 "National Integrated Protected Areas Systems (NIPAS) Act of 1992" stipulates that the management, protection, sustainable development and rehabilitation of protected areas shall be undertaken primarily to ensure the conservation of biological diversity. However, not all of the remaining natural forests are covered by NIPAS. All remaining old-growth forests are protected but logging is still allowed in secondary forests.
- RA No. 8371 "Indigenous People's Rights Act of 1997" recognizes the vested rights of indigenous peoples over their ancestral lands within forestlands including secondary forest.
- Executive Order (EO) 363 of 1995 adopts Community-Based Forest Management (CBFM) as the national strategy to ensure the sustainable development of the country's forests and promote social justice.
- EO 318 of 2004 "Promoting sustainable forest management in the Philippines" attempts to revise PD 705 and aims to attain sustainable forest management in the country's production forests.

All of the above provide the overall framework for climate change adaptation in the Philippines. Watershed management, forest conservation, and greater local community participation could help in climate change adaptation. For example, protecting existing forests allows for natural adjustment to a new climate regime. Greater local community involvement could minimize financial cost of adaptation for state agencies.

In terms of actual ground activities, the government has been actively pursuing several initiatives in spite of its limited resources. These include the following:

- Conservation of remaining forests in NIPAS sites and watershed areas;
- Reforestation and rehabilitation of barren upland areas through tree planting and agroforestry;
- Community-based forestry activities such as community organizing and development.

The private sector is less involved today compared to the height of logging activities in the 1950s and 1960s. However, civil society is more involved as community-based programs increase.

Climate change is hardly being considered at all in the planning process of the government for forest resources. The more urgent concern is to save remaining forests from human exploitation that is the more imminent threat.

As shown earlier (Section 4.2), certain forest types in the Philippines, especially the dry forest types, are highly vulnerable and could be replaced by other types of forests. The laws and regulations mentioned above may need to be re-assessed and updated to focus more on how forest management can be improved to mitigate climate change, with special interest in areas classified as dry forests, such as areas in Northern Luzon, Negros, Cebu, Palawan, Basilan, and General Santos.

In light of this, a national adaptation strategy should probably focus on identifying forested areas that are more at risk and the unique species these areas harbor. Specific adaptation options could include conservation and management of vulnerable species, assisting local communities that are highly dependent on forests at risk, and others.

5.1.2. Adaptation Strategies Identified by Stakeholders in PCW

The following adaptation strategies to mitigate impacts of climate change in PCW ecosystems are discussed below. These are identified by local communities, stakeholders, LGUs, and POs through workshops, FGDs, and interviews. Adaptation strategies identified are based on what kind of ecosystem is being affected by climate change.

Lowland Farms

- Early onset of the rainy season

Carranglan farmers use short term varieties while those from Pantabangan install SWIPs.

- Late onset of rainy season

As lowland farms in Pantabangan are strongly vulnerable to late onset of the rainy season, an adaptation measure done to cope with its impacts is the choice of adaptable species for climate change.

- La Niña

Adaptation measures undertaken in Pantabangan include increased consultation with different concerned agencies and the involvement/participation of different stakeholders.

- El Niño

Farmers in Pantabangan and Carranglan shift to drought resistant crops and apply supplemental watering to adapt to El Niño.

- Rainy season

Lowland farms in Pantabangan are only moderately vulnerable to rainy season, thus, no adaptation measures is being undertaken by the farmers.

Upland Farms

- Early or late onset of rainy season

Participants from Pantabangan and Carranglan mention that during early or late onset of rainy season, they make use of appropriate variety of planting materials, installation of firelines, strict implementation of forest laws, adoption of modern method of farming suited for upland and visibility of enforcement agencies to the area.

- La Niña

Farmers delay their planting activity in response to La Niña.

- El Niño

Participants mention that farmers shift to more tolerant crops during El Niño. For instance, farmers plant crops whose growth are not inhibited by limited supply of water. Thus, traditional crops such as rice are not produced during occurrence of El Niño episode.

- High temperature/Summer season

Participants mention the use of drought-resistant crops as commonly practiced by farmers to adapt to high temperature/summer season.

- Rainy season

Farmers use fungicides/pesticides in their upland farms to control the attacks of fungi.

Tree Plantation

- Early or late onset of rainy season

Adaptation measures being undertaken by communities whenever there is an early or late onset of rainy season are: (1) adjust silvicultural treatment schedules; (2) plant species that can adjust to variable climate situations; and (3) plant trees at the proper time.

- La Niña

An adaptation measure undertaken to overcome effects of La Niña is the implementation of proper silvicultural practices.

- El Niño

Measures being undertaken by concerned sectors to adapt to the effects of El Niño include construction of fire lines to avoid spread of forest fires, practice of controlled burning, and supplemental weeding.

Grassland

- Early onset of the rainy season

Adaptation measures undertaken in Pantabangan include: (1) dependence on forest resources as source of livelihood; (2) increase fund for forest protection, regeneration from national government; and (3) increase linkage among LGUs, government organizations GOs and NGOs.

- La Niña

Adaptation measure undertaken to minimize impacts of La Niña in Carranglan is the introduction of immediate drainage.

- El Niño

Grasslands in Pantabangan and Carranglan are strongly vulnerable to El Niño. To cope with impacts, adaptation measures done are application of controlled burning and introduction of drought-resistant species.

- High temperature/Summer season

To decrease the degree of impacts of high temperature in the grasslands of Pantabangan and Carranglan, intensive information dissemination campaign among the stakeholders is undertaken.

Natural Forests

- Early onset of the rainy season

Adaptation measures undertaken to reduce the impacts of early onset of rainy season on the natural forests are: (1) placing safety net measures for kaingineros by the local and national government; (2) developing strong linkage between the LGU and the government agency to protect the forests; and (3) empowerment of the local communities.

- Late onset of the rainy season

Adaptation measures being implemented to abate the impacts of late onset of rainy season include: (1) urging the kaingineros to apply safety measures; (2) developing strong linkage between the LGU and the government agency to protect the forests; and (3) empowering local communities.

- La Niña

No adaptation measure is mentioned in Pantabangan and Carranglan, however, in Ma. Aurora, a drastic move such as cancellation of the timber license agreement or implementation of total logging ban is being implemented.

- El Niño

Among the municipalities of Pantabangan, Carranglan, and Ma. Aurora, only Pantabangan noted the adaptation measures that are undertaken. These are: (1) safety measures being undertaken by kaingineros; (2) strong linkage between the LGUs and the different government organizations; and (3) empowerment of the local communities.

5.2. Water Resources

5.2.1. Common Adaptation Measures

Workshops and FGDs reveal various adaptation strategies to mitigate climate change impacts and vulnerability of water resources in PCW.

Whenever there is water shortage, farmers commonly resort to shallow tube wells to irrigate their farms (Table 5.1). Others source their water from nearby streams using pumps to bring water to their cultivated fields.

From the list of adaptation measures, switching to alterative crops is a common practice. However, this is done only in selected areas because the soil type is not suitable for other crops aside from rice. Although the production cost in planting alternative crops like vegetables is more expensive than the production cost of rice, some farmers take the risk rather than wait for the next cropping season. Production cost for rice averaged at about PHP 25,000 per hectare while average production cost for vegetable is between PHP 38,000 to PHP 40,000 per hectare depending on the source of irrigation. It is cheaper when irrigation is provided through the UPRIIS than getting water from shallow tube wells or other alternative sources. Nevertheless, this adaptation strategy is very effective since the net income the farmer gets is higher than from the net income from rice production. The yearly net income of the farmers in vegetable production is about PHP 110,000 per hectare since they can harvest at most 12 times in a cropping season while the annual net income in rice production is only PHP 25,000 per hectare on the average.

Table 5.1. Current adaptation to floods and droughts at the UPRIIS, PCW

	District 1	District 2	District 3	District 4
Adaptation Strategies for Drought	Use of shallow tube wells Planting of new varieties of rice (i.e. Gloria rice) and other crops that have less water requirements Rotation method of irrigation is being implemented.	Use of shallow tube wells Planting early maturing varieties of crop and vegetables Scheduling (rotation method) of irrigation is being implemented Use the direct seeding method which requires less water	Use of shallow tube wells Use of other water sources (i.e. from the Atate river that is connected directly to the irrigation main canal) Others are hanging around and relying on the crops of their neighbors	Use of shallow tube wells Use of other water sources (i.e. from the Peñaranda river that is connected directly to the irrigation main canal) Others are hanging around and relying on the crops of their neighbors
Adaptation Strategies for flood	None, merely waits for the next cropping season	Repair the damages Switch to other crops that can sustain floods and heavy rainfall Livelihood diversification (swine production, vegetable farming, canton making and fruit juice making) through Farmers' Business Resource Cooperative Construct fish ponds in the flooded area	Switching to early maturing varieties of crops (i.e. from palay to corn) Attend seminars and trainings conducted by stakeholders about crop production Use of solar dryers	Use of solar dryers

Furthermore, the respondents also acknowledge the role of UPRIIS-NIA in program implementation and physical rehabilitation and repairs of infrastructure like irrigation canals and water pumps, although there are limited funds and resources to cover the expenses. The NIA also offers loans to qualified farmers and use it as incentives to IAs that are able to pay at least 80% of the irrigation service fee. During dry season the farmers pay 3.5 cavans per hectare which is approximately PHP 1,750. During wet season they pay 2.5 cavans per hectare which is approximately PHP 1,125.

The IAs seek assistance from LGUs to provide a multi-purpose pavement for drying palay even during rainy season. Since most of the residents in the service area are farmers, a facility like this is a must for them. Likewise, the acquisition of a solar dryer is another technological adaptation that could help them cope with the threats of floods or extended rainy days that adversely affect their livelihood.

Agencies like DA and LGUs, from time to time, provide free rice seedlings and water pumps to reduce the expenses of farmers in terms of farm inputs. In addition, the municipal agricultural officer (MAO) provides trainings to inform farmers about the latest techniques in farming and hybrid rice seeds that yield more than the ordinary rice varieties. A well-informed community is most of the time the best weapon to cope with the impacts of water shortage and floods.

Among the various facilitators and promoters of adaptation measures, the MAOs have either developed or administered the most number of coping mechanisms. This can be credited to their mandated responsibility to the community, as well as to the strong personal relation they have established with the people in the course of time. Meanwhile, the UPRIIS-NIA is responsible for the maintenance of the irrigation infrastructures. However, NIA is often limited by inadequate funds and resources that reduce their effectiveness in helping reduce the vulnerability of the farmers to water shortage and floods.

5.2.2. Potential Adaptation Strategies

The identification of the adaptation strategies is done through review of relevant documents such as the National Action Plan/National Communication of the Philippines, solicitation of experts' judgment through interviews, participatory workshops involving farmers, local government officials, officials and personnel of the NIA, among other key stakeholders in the watershed. The results of these various activities are analyzed and synthesized and submitted to the stakeholders for validation and approval.

Engagement of the various stakeholders in the development of adaptation strategy is critical not only because much of the strategy will be directly implemented by stakeholders, but also because most of the strategies has wide ranging implications on the general public as well as on many institutions and agencies of the government. The decision to implement a particular set of adaptation measures normally entails significant changes or adjustments in the way people think and do things relative to its natural environment, individually and in relation to one another. The challenging part is how to bring all the stakeholders together to agree on what the real risks related to climate change are and then agreeing further on what steps to take so as to cope with these risks in the most effective and efficient way. Arrival at an agreement on what strategy to implement rests on the facilitation of negotiation between and among the stakeholders whose interests come first and how much cost, if at all, will any stakeholder have to bear once the strategy is implemented.

In PCW the development and implementation of the adaptation strategy is envisioned to be carried out using a three-level mode where the selection, approval and commitment to support the implementation of a strategy heavily involved the local communities, the LGUs, and the DENR, along with other stakeholders. However, due to lack of material time and resources, the project never got to the approval and commitment seeking phase.

Based on the outcomes of the FGDs, key informant interviews, workshops and review of relevant documents, a list of the potentially suitable adaptation measures is made and shown in Table 5.2. Most of the measures listed are commonly practiced by farmers while others are identified as essential measures that must be implemented to cope with the changing climate and, at the same time, promote sustainability of watershed resources and services they provide. Of these measures the most common recourse for farmers is to resort to alternative sources of water, i.e., shallow tube wells and tapping into nearby rivers. The attractiveness of this strategy lies in the opportunity it provides to do farming even with the shortage of water from the irrigation system. The downside to this strategy is the cost involved that cuts significantly into the net revenue of the farmers' produce. It is also worth noting that many of these adaptation measures are applied autonomously by farmers with almost no support or intervention from the government. This could explain why for most of the measures listed, the absence of adequate support services is identified as a common limitation to the adoption of these measures.

Many of the farmers who are currently constrained by the cost of adaptation measures and the limited funds they have are crying out for government intervention to help them carry the extra financial burden of adaptation, mainly by helping them raise the highly regulated price of rice in the market. Under the current pricing scheme for rice, there seems to be little incentive for the farmers to undertake measures that will entail additional production cost to alleviate losses from floods or droughts or simple water shortage when in the final analysis they will still end up losers. Another constraining factor identified is the absence of skills for adaptation. The success of others in adapting to extreme climate-related events can provide motivation for others to do the same. However, the lack of proficiency for adaptation usually hampers farmers from even trying. If ever they do, they are often discouraged by the poor initial results they get. This is where training and skills development programs become useful. But like other adaptation measures training and skills development measures are also constrained by limited funds. The farmers alone will not have the capacity to carry out this measure, though in extremely extraordinary case some farmers may be able to do it. Obviously such adaptation measure will have to be facilitated by technically and financially-able government and private agencies.

Improved watershed management, reforestation and soil erosion control are also identified by farmers and other lowland stakeholders as potential adaptation measures. In relation to the lowland farmers, these measures will not really improve their ability to adapt to floods and water shortage but its successful implementation could minimize the occurrence of floods and water shortage that will reduce farmers' vulnerability.

5.3. Local Communities

5.3.1. Adaptation Strategies

Local communities employ different adaptation strategies for food, water, livelihood, and health.

Food: Availability of Planting Materials

From the 156 (64.20%) respondents who mention that climate events have an effect on availability of planting materials, 96 employ adaptation strategies during variable and extreme climates (Table 5.3).

The most common strategies are improving water system (25%), using fertilizers and pest control (18.75%), waiting for the perfect planting time (13.54%), changing crops (12.50%), buying seeds from the DA/Philippine Rice Research Institute (10.42%), producing their own seeds (10.4%), storing seeds (5.21%), exchanging seeds (3.13%), and planting early (1.04%).

Table 5.2. Potential adaptation measures for the lowland farmers of UPRIIS, PCW

Adaptation	Common Features					
	Costs	Benefits	Technical Viability	Acceptability to Farmers	Support Services	Key Limitations
Drought/Water Shortage						
Shallow tube wells	Medium to high	High	High	High	Inadequate	Lack of capital
Nearby rivers	Medium	High	High	High	Inadequate	Lack of capital
Switching to vegetable farming	High	High	Low to medium	High	Inadequate	"Initial skills and capital, site suitability"
Switching to early maturing crops	High	High	Low to medium	High	Inadequate	Initial skills and capital
Rotation in receiving water	High	High	Low to medium	High		Lost opportunity
No planting	High	High	Low to medium	Low	Inadequate	Lost opportunity
Reliance on neighbors and friends	High	Low	Low to medium	Medium		Lost opportunity
Floods						
Use of solar dyers	Medium to high	Medium	High	Medium	Inadequate	Capital
Switching to early maturing crops	High	High	Low to medium	High	Inadequate	Initial skills and capital
Livelihood diversification	Variable	High	Low to medium	Variable	Inadequate	Initial skills and capital
Wait for next cropping season	High	Low	Low to medium	High	None	Lost opportunity

Adaptation	Common Features					
	Costs	Benefits	Technical Viability	Acceptability to Farmers	Support Services	Key Limitations
Construction of fishponds in flooded area	High	High	High	High	Inadequate	Capital
Training and seminars on crop production	Low	High	High	Medium	Inadequate	Time and interest
Others						
Improved watershed management	High	High	High	Low to medium	Low to medium	Funds
Reforestation	High	Medium	High	Low to medium	Low to medium	Funds
Soil erosion control	High	High	High	Low to medium	Low to medium	Skill and funds

Source: Based on the results of FGDs, workshops, interviews and review of relevant documents.

Table 5.3. Adaptation strategies employed in the absence or lack of planting materials as affected by variable and extreme climates

Strategy	Pantabangan	Carranglan	Ma. Aurora	Alfonso-Castañeda	Total
Use fertilizers	42.86	35.8	50	20	37.21
Use water pump	14.29	19.75	16.67	60	19.38
Plant other crops	14.29	12.35	16.67		12.4
Early planting	8.57	7.41		20	7.75
Use good quality seeds	14.29	11.11	16.67		11.63
Store palay	2.86	6.17			4.65
Close the other tributary of water	2.86	7.41			6.98
TOTAL	100	100	100	100	100

Table 5.4. Adaptation strategies employed to improve crop production during variable and extreme climates

Strategy	Pantabangan	Carranglan	Ma. Aurora	Alfonso-Castañeda	Total
Wise Use	60.49	57.28	63.64	26.67	56.67
Fetch water from neighbors' wells	11.11	14.56	18.18	20	13.81
Install more water pumps from rivers/streams	9.88	9.71	9.09	6.67	9.52
Improve water system mechanism	2.47	5.83		20	5.24
Dug another deep well	3.7	5.83		13.33	5.24
Store water	3.7	3.88		13.33	4.29
Buy bottled water	8.64	0.97			3.81
Plant trees		0.97	9.09		0.95
Scheduling		0.97			0.48
Total	100	100	100	100	100

Food: Crop Production

Table 5.3 also shows that adaptation strategies also differ among the different municipalities under study. In Pantabangan the most common adaptation strategy is fertilizer use (28.13%) which is followed closely by improving the water system (21.88%). Meanwhile, in Carranglan (21.57%), Ma. Aurora (42.86%), and Alfonso-Castañeda (50.00%) improving water system is the most adopted strategy to make planting materials available especially during variable and extreme climates.

Of the 251 respondents who believed that crop production is affected by variable and extreme climate events, only 129 respondents are actually employing strategies to reduce the impact of climate-related events as shown in Table 5.4. Among the strategies adopted by the 129 respondents are: using of more fertilizers (37.21%), using and/or devising water pump (19.38), planting other crops that do not require much water (12.40%), planting early (7.75%), using good quality seeds (11.63%), storing palay seedlings (also known as binhi) (4.65%), and closing other water tributaries to centralize flow of water as well as maximize the volume of water that irrigates the farms (6.98%).

The last strategy mentioned above is practiced by some residents of Carranglan. Meanwhile, the most common strategy of respondents in Pantabangan (42.86%), Carranglan (35.80%) and Ma. Aurora (50.00%) is the use of fertilizers while respondents in Alfonso-Castañeda (60%) use water pumps.

Domestic Water

The majority of the respondents (210) also employ several adaptation strategies whenever water supply for domestic use becomes limited. The most common strategy is using water wisely or conserving water (56.67%). This pertains to recycling used water for other purposes. For instance, laundry water can be used to clean house or water plants. The rest of the respondents employ strategies such as fetching water from neighbour's wells when their own dries up (13.81%), installing more water pumps (9.52%), improving water system through repair or change of damaged hose (5.24%), digging another well (5.24%), storing water (4.29%), buying water (3.81%), planting instead of cutting trees (0.95%), and scheduling of usage (0.48%).

Irrigation Water

The most common adaptation strategies practiced in the watershed are using water pumps and generator (46.15%), cessation of planting or waiting for the rain to come (19.66%), scheduling water distribution (8.55%), digging wells (5.98%), sourcing water directly from the rivers (4.27%), reducing production (2.56%), early planting (2.56%), diverting the flow of water (0.85%) and choosing drought-resistant crops (0.85%) (Table 5.5). To cope with the impacts of El Niño, Pantabangan (37.93%) and Carranglan (64.94%) respondents resorted to using more water pumps and generator to produce water needed for crop production. On the other hand, 40% of the respondents from Ma. Aurora use water pumps and generator while the same percentage did not plant at all. In Alfonso-Castañeda, half of the respondents did not plant at all to avoid the risk of losing.

Table 5.5. Adaptation strategies employed by the respondents to cope with impacts of variable and extreme climates on water requirements for irrigation

Strategy	Pantabangan	Carranglan	Ma. Aurora	Alfonso-Castañeda	Total
Use water pump and generator	37.93	64.94	40	16.67	46.15
No planting/ wait for the rainy season	27.59	15.58		50	19.66
Scheduling of water distribution	13.79	3.9	40	16.67	8.55
Dig wells	3.45	7.79			5.98
Directly gets water from the nearby body of water	13.79		20		4.27
Reduce the production or planting	3.45	2.6			2.56
Early planting		2.6		16.67	2.56
Divert the flow of the water		1.3			0.85
Change crops		1.3			0.85
Total	100	100	100	100	100

Livelihood

Various adaptation strategies (Table 5.6) are also employed by 104 respondents to decrease the negative impacts of variable and extreme climate on their livelihood. More than half of the respondents save their income (65.38%), 27.88% of them borrow money from relatives and others, and 6.73% ask for support from affluent relatives. Although not a major strategy to reduce livelihood vulnerability, PCW residents have a tendency to seek the assistance of relatives to reduce the impacts of variable and extreme climate events.

Health

Among the adaptation strategies being practiced by watershed communities when they experience health problems are eating healthy foods (55.19%), maintaining cleanliness of surroundings (24.07%), taking herbal medicines or consulting with faith healers or albularyo (12.59%), and going to the hospital for check-up and medicines (8.15%) (Table 5.7).

Table 5.6. Adaptation strategies employed for livelihood during variable and extreme climates in the watershed

Strategy	Pantabangan	Carranglan	Ma. Aurora	Alfonso-Castañeda	Total
Save	61.54	77.36	14.29	40	65.38
Loans to relatives	28.21	18.87	85.71	40	27.88
Support of relatives	10.26	3.77		20	6.73
Total	100	100	100	100	100

Table 5.7. Adaptation strategies used for health problems during variable and extreme climates

Strategy	Pantabangan	Carranglan	Ma. Aurora	Alfonso-Castañeda	Total
Eat healthy foods	51.96	55.07	50	81.25	55.19
Cleanliness	24.51	26.09	21.43	6.25	24.07
Herbal Medicines / Faith healer	10.78	13.77	21.43	6.25	12.59
Go to hospitals for medicines	12.75	5.07	7.14	6.25	8.15
Total	100	100	100	100	100

5.3.2. Analysis of Adaptation Strategies Employed by Communities

An examination of the adaptation strategies practiced by the households is important in assessing their vulnerability to climate variability and extremes. Adaptation strategies pertain to strategies/mechanisms that communities employ to minimize or reduce the impacts of climate variability and extremes. Households tend to be more vulnerable when the adaptation strategy being employed is not effective. Also, since time, money and effort are also needed in many adaptation strategies, households adversely affected by climate variability and extremes become three times more vulnerable if such coping mechanisms prove to be ineffective.

For each of the component index, i.e., food, water, livelihood and health, each household employs different adaptation strategies. Results of the survey revealed that the respondents employed a maximum of eight adaptation strategies whenever households' food, water, livelihood, and health are at risk. Table 5.8 shows the coping mechanisms employed by the communities totaling 1,581 (multiple responses). Almost 18% of the respondents practiced reduction in terms of consumption, i.e., food and water. About 14% availed of loans/credit to increase farm inputs, particularly fertilizer and water generation, which is an added cost for the farmers. Although, it is hard to prove that praying reduces vulnerability, still 13% of the respondents believe that divine guidance helped them cope with all kinds of problems, may it be climate-related or not. Other strategies employed are: storing food, firewood, medicine, and water; community and kinship ties; off-farm work; government/NGO assistance; crop diversification; asset disposal; treebelts/wind breaks/hedgerows; resettlement/ rehabilitation; and the ability to forecast natural hazards/disasters based on community's/indigenous traditional knowledge.

A closer look at the municipalities reveal that the respondents in Pantabangan (75.91%), Carranglan (72.22%), and Ma. Aurora (94.12%) reduce their consumption, except in Alfonso-Castañeda where treebelts/ windbreaks/hedgerows (86.36%) and storing food, firewood, medicine and water (86.36%) are the most commonly practiced strategies. The town of Alfonso-Castañeda faces the sea; hence, windbreaks and treebelts are established to break the intensity of wind. In the three towns, this is not a problem since they are surrounded by mountains and forests. In the towns of Pantabangan and Carranglan, only a small number of respondents are employing the strategy on treebelts/windbreaks/hedgerows, with only 26.28% and 18.18%, respectively (Table 5.8). This is because the residents in these towns are cutting trees for fuelwood, which is also the reason why there are only few trees in these towns.

Availing of loans, as a strategy, is also popular for the respondents (Table 5.8). At least more than half of the respondents in the four towns avail of loans/credit. It is also important to note that this strategy, as seen by the participants of the FGD, is not effective, especially among small farmers. According to them, this practice even made them more vulnerable.

At least half of the respondents in the four towns seek help from their community and kin. Relatives, friends and neighbors serve as the support group, from which the respondents can turn into during times of climate variability and extremes.

Table 5.9 shows that involvement in off-farm work is also a common strategy, which is practiced by 30% of the respondents. Most of the respondents look for work outside their town as factory worker, saleslady, housemaid, construction worker, among others.

Institutions

Table 5.10 summarizes the various adaptation strategies employed by the different institutions in order to adapt to the impacts of climate variability and extremes.

Among the institutions studied, LGU has the most number of adaptation strategies with 23, which can be attributed to their accountability as the one responsible for the whole community and who really associates personally with the people. This is followed by the DENR with 11, then by NIA and NPC both with four. It can be noticed that reforestation/tree planting is very common adaptation strategy among all institutions. They all recognize the importance of trees in water generation and in averting global warming because of the capacity to be a carbon sink. Furthermore, they also recognize the role of trees in controlling floods whenever typhoon comes. In some cases, however, reforestation is not included in the cultural practices of the people, like the Igorots who prefer to plant vegetables rather than trees.

Forest protection activities such as the formation of fire brigade and hiring of forest guards are also practiced as forms of adaptation strategies. These help in controlling and preventing any forest fires due to kaingin and charcoal making, and in the denudation of forest due to illegal logging.

Table 5.8. General adaptation strategies employed by PCW communities during variable and extreme climates

Strategies	Pantabangan (N=137)		Carranglan (N=198)		Ma. Aurora (N=17)		Alfonso-Castañeda (N=22)		Total (N=374)	
	Freq*	%	Freq*	%	Freq*	%	Freq*	%	Freq*	%
Reduced consumption	104	75.91	143	72.22	16	94.12	15	68.18	278	74.33
Loans/credit availability	79	57.66	119	60.1	10	58.82	13	59.09	219	58.56
Pray or make offerings to Anito	88	64.23	126	63.64	10	58.82	16	72.73	204	54.55
Store food, firewood, medicine	70	51.09	120	60.61	7	41.18	19	86.36	198	52.94
and water										
Community and kinship ties	72	52.55	86	43.43	7	41.18	11	50	176	47.06
Off-farm work	49	35.77	60	30.3	5	29.41	8	36.36	122	32.62
Government/ NGO assistance	33	24.09	71	35.86	7	41.18	6	27.27	113	30.21
Crop diversification	39	28.47	42	21.21	6	35.29	4	18.18	91	24.33
Asset disposal	33	24.09	44	22.22	2	11.76	5	22.73	84	22.46
Treebelts/ wind breaks/	36	26.28	36	18.18	2	11.76	19	86.36	59	15.78
hedgerows										
Resettlement/ rehabilitation	7	5.11	14	7.07			7	31.82	18	4.81
	Freq*	%	Freq*	%	Freq*	%	Freq*	%	Freq*	%
Ability to forecast natural hazards/disasters based from community's/ indigenous traditional knowledge	16	11.68	9	4.55			8	36.36	19	5.08

Table 5.9. Adaptation strategies of different user-institutions to minimize the negative impacts of climate variability and extremes

Institutions	Adaptation strategies
NIA	Reforestation Forest protection (campaign) Physical rehabilitation Release of excess water from the reservoir/dam especially in rainy season when the dam is overflowing to avoid flooding
DENR	Reforestation Forest protection (fire brigade) Adjustment in schedule of program implementation/prioritization Monitoring Shading of seedling in reforestation sites Deploying of forest guards to patrol the forest Planting of fire breaks IEC Hiring of additional manpower especially of casual laborers Adjustment in program prioritization Integrated Social Forestry Program (Gen. Luna)
NPC	Reforestation IEC Proper choice of species Adjustment in schedule and implementation

Institutions	Adaptation strategies
LGU	Tree planting and reforestation (Conversion) Provision of relief goods (Bunga and FC Otic)) EIC, information dissemination and conduct of seminar on proper farming, and house visits especially during typhoon season (Villarica, East Poblacion, and Bantug) Creation of task force El Niño/La Niña, formation of disaster brigade (Fatima) Planned action to be taken to avert current and destructive effects of El Niño and La Niña Diversion of program to other urgent problems (Pantabangan), and adjustment in program schedules Bridge development (Abuyo), hanging bridge construction (San Agustin) Repair, development, construction and maintenance of roads (including DPWH) Development and road construction (San Agustin) Provision of solar dryer, multi-purpose pavement for drying palay, (Bunga, Galintuja, Ma. Aurora,) Free medicines (G.S. Rosario) Hiring of extension worker especially BHW (Fatima and Galintuja, Ma. Aurora) Request for free seedlings from DA Helping people in evacuation centers Lending of seedlings (Marikit Repairs of deep well and canals Rehabilitation of destroyed infrastructure and reporting it to the Department of Public Works and Highways (DPWH) (San Juan) Relay information about the status of the barangay after a calamity to the municipal government (Bunga) Barangay tanods visit the community to help with their problems (Bantug) Small organization that buy palay in higher prices and sell rice in lower prices (Salazar), training of people's organization (Lublub) Digging possible water sources for irrigation, , spring or deep diverting the flow of water from the river into the direction of the people farm fields (Bantug) Diverting the flow of water from the river into the direction of the people farm fields through the help of barangay tanod (Bantug) Spring development, formulation and implementation of SWIP, provision of additional water tank, pump and hose, finding other sources water sources, deep well construction and dissemination of water distribution schedules.

Table 5.10. Analytical matrix of cross sectoral impacts (forest/agriculture to water, institutions and local communities)

Adaptation strategy for forests and agriculture	Effect on water resources	Effect on institutions	Effect on local communities
Use of early maturing crops	+ Lower water demand	0	+ Higher income
Use of drought resistant crops	+ Lower water demand	0	+ Higher income
Supplemental watering	- Higher demand for water	- Increase cost of developing alternative sources of water	- Greater labor demand + Higher income
Proper scheduling of planting	0	- Increase cost for training, technical assistance, RandD	0

Table 5.10 (Continued)

Soil and water conservation measures	+ Conservation of water	- Increase cost for training, technical assistance, RandD	- Cash expenses
Establishment of fire lines	+ More vegetative cover promotes good hydrology	+ Less expense on fire fighting	- More labor demand + Less damage to crops from fire; more income
Construction of drainage structures	+ Better water quality (less sediment load)	- Increase cost of implementation	+ Less soil erosion in the farm; greater yield
Controlled burning	+ Less damage to watershed cover	0	0
Tree planting	+ Better hydrology	- Increase cost of implementation	+ Steady supply of fuelwood - Less area for farm
Enhance community-based organizations	+	+ Better participation in the political process	+ Better participation
Total logging ban	+ More forest cover	- Increase cost of enforcement and protection	- Less income - Less sources of income
Use of appropriate silvicultural practices	+/- Could promote or impair hydrology depending on the practice	- Increase cost of implementation	- Increase cost of implementation
Better coordination between LGUs	+ Promotes better watershed management	+ Greater collaboration among LGUs	+ Better delivery of services to farmers
Information campaign	+	+ Increase awareness and competence	+ Increase awareness and competence
Better implementation of forest laws	+ Promotes better watershed management	- Increase cost of implementation	+/- Could adversely current livelihood of farmers that are deemed 'illegal'

Legend: (+) positive impact, (–) negative impact, (0) no effect, (na) not applicable.

These are also connected to the monitoring activities conducting. However, most of the time the number of forest guards in the area is insufficient, hence, they are limited in guarding the forest.

Another adaptation strategy mentioned is the adjustment in program implementation and prioritization. This involves, for instance, the implementation of programs planned and scheduled to be operated and executed for the 2nd quarter but is immediately implemented in the 1st quarter, hoping that it would minimize or solve pressing problems experienced in the area. It also includes formulation of new programs of action to minimize the negative impacts of climate variability and extreme. Among the programs implemented are planting of fire breaks for forest fires, livelihood opportunities for jobless people, SWIP and spring development to help the communities to have a continuous supply of water even if there is El Niño. In time of El Niño/La Niña a task force is also created to help and coordinate with the people.

Additional manpower/workforce is also needed for the proposed program to be implemented. In connection with this, additional money/budget is also needed. For institutions like NIA, DENR, and NPC, the budget is already fixed for a certain year and comes from the central office. Hence, it is very difficult for them to have extra budget for the wages of their employees and in acquiring instruments and materials for the implementation of the programs.

Physical rehabilitation and repairs are also mentioned. These involve the repairs of properties and infrastructures destroyed by typhoons, such as bridges and roads used by the community to travel and transport their products to the market, transmission lines of NPC and construction and improvement of deep wells, water pumps and canals (in the case of NIA and LGU).

The distribution of relief goods is also a common practice in the area. Among the goods given include rice and medicines. However, this may result to dependency especially if the distribution is in a dole out manner. Others are asking the Department of Social Welfare and Development to help in sourcing and gathering of relief goods.

The construction of multi-purpose pavement for drying palay is a strategy to help the farmers dry the palay even in rainy seasons. Region III, wherein PCW is located, is the rice granary of the country. Most of the residents in the region are farmers. Likewise, the acquisition of a solar dryer will also greatly help the farmers especially during the rainy season.

Another adaptation strategy mentioned is asking assistance from different institutions of the government, such as DA for free seedlings of palay, which lessens the expenses of the farmers in terms of farm inputs. They also request water pumps from the same department.

Meanwhile, some barangays like Salazar are encouraged and initiate the formation of small organizations that buy palay with higher prices and sell rice in lower prices.

Information dissemination is also an adaptation strategy practiced by the institutions. Many BDCs disseminate information about coming climate variability and extremes. These enable communities to prepare themselves by obtaining the necessary food and clothing to help them survive the upcoming catastrophic event. BDCs are also concerned about the farming technologies and practices of residents; hence, they also conduct seminars on these matters. Also done are house-to-house visits to see the conditions of the residents.

From these examples, it can be observed that the general function of BDCs is to relay messages of the communities to their respective LGUs. Then the LGU either makes an action, provides solution to the reported problem, or passes it to concerned government agencies such as DA, DENR and Department of Health.

Information, education, and communication (IEC) are also identified as adaptation strategies. The institutions give advance information about the coming climate events (effects), educate and train the people in proper farming that will give them high profit, ask for the feedbacks on certain projects implemented from the community, and identify the needs of the community.

It also important to note that two respondents (Maglanoc and R.A. Padilla) from the BDCs say that they do not employ any strategy to help the communities in adapting to climate variability and extremes because the communities themselves have already adapted to and survived these events.

5.4. Tradeoff Analysis of Adaptation Options

The main hypothesis is that specific sectoral adaptation strategies may complement or conflict with adaptation strategies for other sectors. For example, one of the adaptation strategies in the forest resources sector is reforestation or planting of forest trees in denuded areas. Such a strategy could lead to lower water yield adversely affecting water supply for power generation and irrigation. At the same time, tree planting activities may have positive effects on local communities (e.g., greater fuel wood supply). In this section, we conduct a cross-sectoral analysis of recommended adaptation strategies to determine potential synergies and conflicts between them. An attempt is made to identify 'best bet' and 'win-win' adaptation strategies based on the foregoing analysis.

The concept of tradeoffs is basic to economics and arises from the idea that resources are scarce. As a general principle, tradeoff analysis shows that for a given set of resources and technology, to obtain more of a desirable outcome of a system, less of another desirable outcome is obtained (Stoorvogel et al. 2004a). While there can be win-win outcomes in two dimensions, even such a win-win must come at the expense of some other desired attribute.

Trade off analysis has been used in exploring the effects of change in land use, policies, and scenarios in agricultural production systems in Ecuador and Peru (Antle et al. 2003, Stoorvogel et al. 2004b). In these studies, the researchers developed a simulation model called TOA (for tradeoff analysis) which is a tool for an integrated analysis of tradeoffs between economic and environmental indicators. The analysis to quantify these relationships is based on a multi-disciplinary approach and used biophysical as well as econometric-process simulation models. It is also based on spatially explicit econometric simulation models linked to spatially-referenced biophysical simulation models to simulate land use and input use decisions.

Another form of tradeoff analysis is used in marine protected area management in West Indies (Brown et al. 2000). The approach developed in this study highlighted the objective of enabling decision makers to consider tradeoffs between different criteria to evaluate alternative management options.

Here, a less technically demanding approach is proposed in analyzing the tradeoffs between adaptation options in various sectors which policy makers and stakeholders can use as a first order estimate. Tradeoffs are analyzed using matrices that show the positive and negative interactions between sectoral adaptation options. The positive and negative ratings are based on the researcher's judgment. Mitigation measures are identified to minimize or eliminate adverse impacts. Adaptation strategies common to all sectors are also identified.

5.4.1. Effects of Adaptation Strategies for Forests and Agriculture

The effects of adaptation strategies for forests/agriculture on water resources are generally positive (Table 5.10). This suggests a very synergistic relationship between adaptation for forest/agriculture and water. In contrast, adaptation strategies for forests/agriculture have mixed effect on the various institutions in PCW. Most of the adaptation strategies recommended require additional investments. Under tight budget constraints of many Philippine agencies, this could pose a significant hurdle to the implementation of the recommended strategies. Likewise, the effects on local communities are mixed. In some cases there are positive effects and quite the opposite in others. For example, it is possible for farmers to obtain higher yield and income as a result of adaptation options such as the use of appropriate crop varieties. However, some adaptation activities such as supplemental watering could require more labor.

Adaptation strategies for the forest/agriculture sector could be prioritized based on their effects on other sectors (in addition to their effectiveness in forestry/agriculture). In general, those that have positive effects on other sectors should receive higher priority while those which have negative effects could be mitigated, if possible.

Among technical adaptation strategies, the use of early maturing crops and drought-resistant crops have the most positive effects to other sectors. The establishment of fire lines has also a general positive effect but requires more labor time to establish. Similarly, social adaptation strategies (e.g. community organizing) have positive effects on the other sectors.

For most of the adaptation strategies there are clearly tradeoffs such that there are both positive and negative effects on other sectors. In many cases, the negative effect is the additional cost required to be able to implement adaptation strategies. This hurdle may prove daunting considering the lack of resources of many Philippine government and non-government agencies.

5.4.2. Effects of Adaptation Strategies for Water Resources

Adaptation strategies for water resources have overwhelmingly positive effects on forest and agricultural crop production (Table 5.12). This is understandable considering that proper water management is essential to crop growth and development. Consequently, farmers are expected to have greater income if adaptation strategies for water are implemented. On the other hand, the effect on institutions is mainly negative in the sense that they will incur more expenditures since many of the adaptation strategies require some reconstruction work, engineering, retooling, increased provision of training, technical assistance, and other related services. This implies that in the face of limited financial resources, adaptation strategies for water may not be implemented fully.

Table 5.11. Analytical matrix of cross sectoral impacts (water to forest/agriculture, institutions and local communities)

Adaptation strategy for water resources	Effect on forest resources/ agriculture	Effect on institutions	Effect on local communities
Reforestation/AF farming	+ Greater tree cover	- Higher investment cost	+ More income
Soil and water conservation measures	+ Increased yield	- Higher investment cost	+ More income
Water impoundment	+ Increased yield	- Greater expenses	+ More income - Greater expenses
Well construction	+ Increased yield	- Greater expenses	+ More income - Greater expenses
Cloud seeding	+ Increased yield	- Greater expenses	+ More income
Use of appropriate crops/ varieties	+ Increased yield	- Greater expenses for RandD, TA, IEC	+ More income
Irrigation management	+ Increased yield	- Greater expenses for implementation	+ Increased income
Tap other water sources (e.g. rivers)	+ Increased yield	- Greater expenses	+ Increased income
Fishponds in flooded areas	+Decreased pressure on forests and agricultural resources	- Additional expenses for TA	+ Increased income - Greater expenses
Repair of damaged infrastructure	0	- Greater expenses	0
Shift in livelihood	+ Less use of land	- Additional expenses for TA, training	+ Increased income
Strict implementation of forest laws	- Could affect crop production in areas deemed for forest	+ Strengthen role of regulatory agencies	+/- Promote peace but possibly lower income
Research on ground water	0	- Greater expenses for RandD, TA, IEC	0
Capacity building activities	+ Build up of mass of competent players	- Greater expenses for RandD, TA, IEC	+ Build up of mass of competent players

Legend: (+) positive impact, (–) negative impact, (0) no effect, (na) not applicable.

Table 5.12. Analytical matrix of cross sectoral impacts (institutional to forest/agriculture, water, and local communities)

Adaptation strategy for institutions	Effect on forest resources/ agriculture	Effect on water resources	Effect on local communities
Reforestation	+ Increased tree cover	+ Better watershed cover	+ Source of fuelwood/ tree products
Forest protection	+ Reduce forest destruction	+ Better watershed cover	+/- Could affect source of forest products
Physical rehabilitation	0	0	+ Better facilities
Release of water from the dam	0	0	- Flooding in low lying areas
Adjustment of schedule	0	0	0
Fire break establishment	+ Reduce fire loss	+ Better watershed cover	+ Reduced fire loss
Community based management	+ Better forest land management	+ Better watershed cover	+ Empowerment of local people
Development of water sources	+ Increased crop yield	+ Stable water supply	+ Increased crop yield
Hiring additional personnel	+ Better forest protection	+ Improved water quality and regimen	0+ Additional sources of income
Proper choice of species	+ Increased yield	0	+ Increased income
Provision of relief goods	+ Reduction of pressure on forest and agricultural resources	0+ Reduction of pressure on water	+ Relief goods supplied
Creation of task forces	+ Better coordination	+ Better coordination	+ Better coordination
Infrastructure repair and construction	+ Increased farm yield	+ Stable and more efficient water supply	+ Increased income
Information, education and communication	+	+	+
Training of PO	+ Better forest/farm management	+ Better water management	+ Skills developed

Legend: (+) positive impact, (–) negative impact, (0) no effect, (na) not applicable.

One possible approach is to determine the costs of recommended adaptation options to know which are affordable given current budget constraints of those who will implement them.

5.4.3. Effects of Adaptation Strategies for Institutions

Adaptation strategies identified by various institutions have generally positive effects on the other sectors (Table 5.12). This shows that the identified strategies are holistic in nature. Of course, the underlying assumption is that financial resources are available to implement these strategies which is the major constraint identified above. There are a couple of exceptions to this. First, stricter enforcement of forest protection rules could adversely affect farmers with no clear land tenure instruments. Many farmers in PCW are informal settlers and forest protection officers could compel them to leave their farms. Second, the release of water in the Pantabangan Dam to prevent overflow could lead to flooding in downstream communities.

Overall, most of the adaptation strategies identified for all sectors have mixed effects (both positive and negative). This suggests that in most cases, adaptation strategies are not neutral, i.e., they could affect other sectors both positively and negatively. Thus, a cross-sectoral analysis should be done at the watershed scale to ensure that negative effects are anticipated and mitigated before the implementation of adaptation strategies.

5.4.4. Common Adaptation Strategies among All Sectors

Three adaptation strategies are common to all the sectors: tree planting/reforestation, selection of appropriate species/crops, and better implementation of laws (Table 5.13). The results reflect the high degree of consciousness of stakeholders on the importance of forests in the watersheds. A number of adaptation strategies are also identified by two sectors. For example, the use of soil and water conservation strategies is explicitly identified in the forest and water sectors and is also implied in the institutional sector. These suggest that individual adaptation strategies could address more than one sector allowing for greater synergy and cost-efficiency.

Table 5.13. Degree of similarities in adaptation strategies among all sectors

Adaptation strategy	Forest / Agriculture	Water	Institutions
Tree planting/ reforestation	X	X	X
Selection of appropriate crops/varieties	X	X	X
Better implementation of forest laws	X	X	X

Adaptation strategy	Forest / Agriculture	Water	Institutions
Soil and water conservation measures	X	X	
Establishment of fire lines	X		X
Construction of drainage	X		
Controlled burning	X		
Enhance community-based organizations	X		X
Total logging ban	X		
Appropriate silvicultural practices	X	X	
Better coordination between LGUs	X	X	
Information campaign	X	X	X
Water impoundment		X	
Well construction		X	
Irrigation management		X	
Cloud seeding		X	
Develop other water sources		X	X
Research		X	
Capacity building		X	X
Release of water from the dam			X
Adjustment of schedule			X
Additional personnel			X
Provision of relief goods			X
Creation of task forces			X
Infrastructure repair and construction			X

5.4.5. Quantification of Tradeoffs

The approach used in this study is qualitative. Succeeding studies may wish to quantify tradeoffs in adaptation strategies between sectors. Some indications of the financial tradeoffs are discussed below.

While reforestation has been identified as a desirable adaptation strategy by all sectors, the tradeoff to institutions which will finance it may be quite high. In the Philippines the official cost estimates of DENR for reforestation is about USD 900 per ha for three years (DENR 1999). Such level of investment may prove limiting for those organizations which would bear the cost of reforestation such as NIA, NPC, and DENR. One possible mitigating measure is to explore more community participation to lower the labor cost since it is the biggest fraction (about 70%) of the total cost of reforestation.

Another adaptation strategy identified by stakeholders as part of forest/grassland fire prevention in the watershed is fire line construction. This strategy entails less cost than reforestation (about USD 20 per ha). Although it is typically part of reforestation work, it can be constructed separately.

Other adaptation options have even less cost. For example, proper scheduling of planting to coincide with the late or early onset of the rainy season involves practically no financial cost to farmers and institutions. And yet the effect of farm yield and income could be very high.

5.4.6. Presenting Results to Policy Makers and Local Stakeholders

One way of visually presenting the results of the study in a nutshell to policy makers is shown in Table 5.14. The chart summarizes the impacts of adaptation strategies in one sector to other sectors. In this way, the potential synergies and conflicts are immediately captured. For example, adaptation measures in water resources have mostly positive effects on forests and agriculture. These include tree planting and provision of irrigation water which directly benefit forestry and agriculture. However, adaptation in water resources will have a largely negative effect on institutions. This can be attributed to the high cost of many of the adaptation measures identified such as construction of shallow tube wells and impounding structures. On the other hand, the effect of adaptation measures in forest resources and agriculture to local communities is mixed, i.e., some are positive while some are negative. For example, use of more resistant varieties could lead to higher incomes but establishment of fires lines mean higher labor cost.

Through this matrix, policy makers and stakeholders will be able to flag which adaptation measures need special attention and further study. Those with negative interactions could also be prioritized for quantitative analysis of tradeoffs.

Table 5.14. Summary of effects of adaptation strategies in one sector to other sectors

	Forest/ Agriculture	Water	Institutions	Local communities
Forest/ Agriculture	na	+	+ -	+ -
Water	+	na	-	+
Institutions	+	+	na	+

5.4.7. Management And Policy Implications

The foregoing discussion has several management and policy implications in the design and implementation of climate change adaptation strategies in Philippine watersheds.

1. Adaptation strategies in one sector could have positive and/or negative impact in other sectors, that is, tradeoffs do exist. This implies that while sectoral analyses have their merits, they are not sufficient. A cross-sectoral analysis at the watershed scale should be done to reveal potential synergies and conflicts between sectors.
2. Cross-sectoral analysis of adaptation strategies will enable managers to anticipate potential conflicts early on. As we have shown, certain adaptation strategies could negatively affect other sectors. For example, reforestation may require more labor from farmers or increased expenditures by government agencies. If these effects are not considered, adaptation strategies may not be implemented at all for lack of cooperation by affected sectors. By considering these at the beginning, there will be greater opportunities for finding solutions.
3. It is possible to identify climate change adaptation strategies that could address more than one sector, thus enhancing synergy. A good example of this is tree planting/reforestation which is identified as an adaptation strategy by all three sectors. By focusing on such strategies, conflicts are avoided. There is also a greater chance of stakeholder acceptance when all are convinced of the desirability of implementing common adaptation strategies.
4. Cost is the major limiting factor of adaptation strategies. The most common tradeoff identified for all sectors is the additional cost the will be incurred in the implementation of adaptation strategies such as in the construction of water impounding structure or in tree planting. In developing countries such as the Philippines, priority for climate change adaptation is low. Adaptation strategies that meet other 'more important' goals may have better chances of implementation. For example, reforestation and tree planting are on-going in the watershed irrespective of climate change considerations.

Box 3. Summary for decision makers

Climate change may have positive or negative impacts to different forest ecosystems in PCW. The degree of impacts of climate change varies depending on the type and location of forest ecosystems being affected. La Niña and El Niño have the most negative impacts to forest ecosystems. Lowland farms, grasslands, and natural forests are the most vulnerable among the forest ecosystems.

In the light of this, an overall adaptation strategy should probably focus on identifying which forest ecosystem in the watershed is more at risk. Specific adaptation options could include assisting local communities shift from forest products from forests at risk, and so on.

An ideal program for reducing the vulnerability of farmers should combine improvement of the conditions of the denuded watersheds of PCW, improvement of the physical and administrative infrastructure of UPRIIS, and enhancement of the ability of the farmers to cope with floods and water shortages. To enhance the coping capacity of the farmers in the service areas of UPRIIS, there is a need to increase the ability of the farmers: (1) to gain access to cheaper alternative water sources; (2) to engage in alternative cropping systems and viable alternative livelihoods; and (3) to set in place systematic and deliberate mechanisms for providing technical and other logistical assistance to the farmers particularly designed to increase and sustain adoption of appropriate adaptation strategies.

The institutions in PCW are faced with varying impacts from the occurrences of climate variability and extremes. Though these climate events both present positive and negative effects, the adverse impacts outweigh the beneficial ones. These affect the operation, budget requirement, workforce, program implementation, and the delivery of services of these institutions. Particularly, these events affect the major services provided by the watershed, i.e., irrigation and power generation.

Among the major strategies developed by the institutions to cope with the negative impacts of climate change are reforestation, forest protection, IEC, and adjustments in the implementation of programs. At the community level, several technological adaptations are also performed such as construction of multi-purpose pavement for drying palay, acquisition of a solar dryer, SWIP, and spring development. The LGUs also have the most number of adaptation strategies developed.

Advance information on the occurrence of severe climate events is the support mostly received by the communities from the institutions. However, a great portion of this comes from NGOs. Nevertheless, the communities are highly satisfied (80%) with the support they receive.

While most negative impacts of climate variability and extremes affect the various aspects of operation of institutions in PCW, institutional adaptation strategies focus more on reforestation and forest protection. Although these play an integral and critical part in their operation, particularly in improving the condition of the watershed, adaptation measures that will address the needs of the institutions themselves should also be developed in order to provide efficient service to its clients. This could include procurement of new equipment, construction of facilities, and adoption of relevant technologies that could enhance their mechanism to cope with the adverse impacts of severe climate events. Obviously, availability of funds is a major hindrance in the implementation of these strategies. Alternative financial sources should be explored to support the other activities of the institutions in order to successfully adapt to the impacts of climate variability and extremes.

Chapter 6

LAND USE CHANGE AND CARBON BUDGETS IN PANTABANGAN-CARRANGLAN WATERSHED

Tropical forests are undergoing massive land cover and land use changes. In the 1990s global deforestation rate of humid tropical rainforests was estimated at 5.8 ± 1.4 M ha (Archard et al. 2002). Between 1990–2000 the estimate was about 8.6 M ha (Mayaux et al. 2005). According to the Millennium Ecosystems Assessment (2005), the main drivers of change in tropical forest ecosystems are habitat change and over-exploitation, and the trend is getting worse. Specifically, the direct causes of tropical deforestation are: agricultural expansion, wood extraction, and infrastructure expansion, while the underlying causes of deforestation include: macroeconomic factors (e.g. trade policies), governance factors (e.g. property rights), cultural factors and demographic factors (Kanninen et al. 2007).

The Philippines was no exception to this trend. In the year 1900, there were still 70% or 21 M ha of forest cover (Garrity et al. 1993, Liu et al. 1993). However, by 1996 there were only 6.1 M ha (20%) of forest remaining (FMB 1998). Thus, in this century alone, the Philippines lost 14.9 M ha of tropical forests. The direct and indirect causes of deforestation include shifting cultivation, permanent agriculture, ranching, logging, fuelwood gathering, and charcoal making (Kummer 1992).

As a result Philippine forest ecosystems have been historically a source of CO2. From the 1500s to the modern era, it is estimated that deforestation has contributed 3.7 Gt C to the atmosphere (Lasco 1998). Of this amount, 70% (2.6 Gt) was released this century alone. However, present land-use cover also absorbs carbon through regenerating forests and planted trees. The vast areas of degraded land in the Philippines in fact offer great potential for carbon sequestration through rehabilitation activities such as reforestation and agroforestry.

This chapter modeled land use change in PCW and the carbon budgets of the PCW ecosystems. The land use change scenarios were used in assessing impacts of climate change on water resources. The carbon budget was determined to assess the potential contribution of PCW to climate change mitigation.

The PCW is host to a range of land uses from natural forest, grassland, lake, rice paddies, reforestation areas and A and D lands.

Because of its high spatial resolution, CLUE-S Model, or the regional extent is used for the site. Figure 6.1 shows the overview of the information flow of the CLUE-S Model (Verburg and Veldkamp 2004). There are four kinds of requirements in order to run the model: land use requirements, land use type specific conversion settings, location characteristics, and spatial policies and restrictions. Land use requirements are computed independently based on the extrapolation methods. Spatial policies and restrictions indicate areas that are restricted to land use change. Allowed land use changes and spatial policies are specified in the matrix conversion. Land use type specific conversion settings determine the flexibility of the land use types to be converted to other land use types and vice versa. Location characteristics determine the suitability of the different land use types to various biophysical and socioeconomic factors.

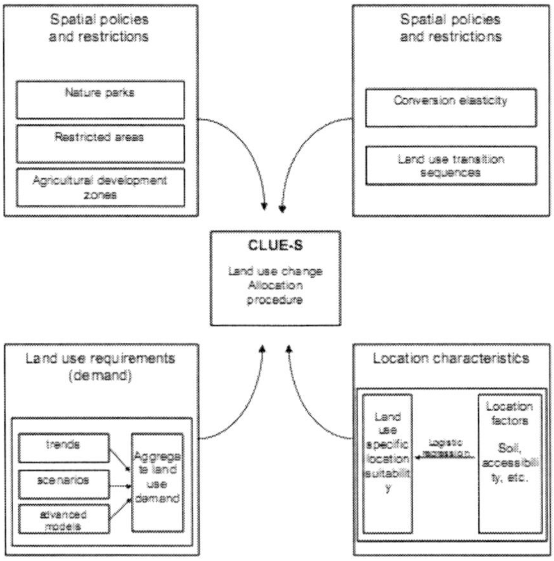

Figure 6.1. Overview of the CLUE-S model information flow.

6.1. Model Parameterization

Land Use Data

Land use map is obtained from the Environmental Remote Sensing and Geo-Information Laboratory (ERSG) at the College of Forestry and Natural Resources, University of the Philippines Los Baños. Since the available map is dated during the early 1990s, a more recent land use map (1999) is gathered from the Provincial Environment and Natural Resources office at Cabanatuan, Nueva Ecija at a scale of 1:50,000. Since there is no current land use map available for the provinces of Nueva Vizcaya and Aurora, it is assumed that land use data are still valid. Table 6.1 shows the existing land use for PCW.

Location Factors

Several location factors gathered from ERSG are considered to have an effect on the spatial variability of the land use types. Biophysical data such as topographic maps: aspect, slope, and elevation maps, and soil maps are used to represent the suitability of the watershed to the different land use types. Distances to streams and roads, and the population density are used to represent the accessibility of the watershed.

Table 6.1. Land use types in PCW

Land use type	Description	Source
A and D / Paddy	Alienable and disposable Land	1:50,000 Land Use Map Environmental Remote Sensing and Geo-Information Laboratory, College of Forestry and Natural Resources, University of the Philippines Los Baños and Provincial Environment and Natural Resources Office, Cabanatuan, Nueva Ecija
Brushwood	Brushwood	
Cultivated ISF	Cultivated integrated social forestry	
Forest	Forest trees	
Grass	Grassland	
Lake	Pantabangan Dam	
Reforestation	Plantations	

In order to test the suitability of the mentioned location factors, a logit model using the ROC Method is used. Land use types are used as the independent variables and the location factors for the dependent variables. ROC values ranges from 0.5 (completely random) to 1 (perfectly fit). For most of the land use types, a reasonable fit is obtained. The result of the ROC Method (Table 6.2) illustrates the location factors affecting the specific land use cover. It can be noticed that only grassland has a relatively low ROC value.

Conversion Elasticity

Low conversion elasticity is assigned to brushwood and grassland, since both are considered to be highly dynamic and can easily be converted to other land use types. This is based on the observation that grassland can be easily cultivated or planted with forest trees. However, high conversion elasticity is assumed for alienable and disposable land, cultivated ISF, and lake since these cover types are highly inelastic. As for forest and reforestation, it is assumed that both cover types have a medium conversion elasticity.

Table 6.2. ROC results

Land use types	Variables	ROC
A and D	Elevation	0.912
	Slope	
	Distance to road	
	Distance to stream	
	Population density	
	Soil: ACL	
	Soil: ACSL	
Brushwood	Elevation	0.89
	Distance to road	
	Distance to stream	
Cultivated ISF	Elevation	0.871
	Aspect	
	Distance to stream	
	Soil: ACL	
Forest	Elevation	0.912
	Slope	
	Aspect	
	Distance to road	
	Distance to stream	
	Population Density	
	Soil: ACL	
	Soil: MSU	
Grassland	Elevation	0.674

	Aspect	
	Distance to road	
	Distance to stream	
	Population Density	
	Soil: ACL	
	Soil: MSU	
Lake	Elevation	0.993
	Slope	
	Distance to road	
	Distance to stream	
	Soil: MSU	
	Soil: US	
Reforestation	Elevation	0.848
	Slope	
	Distance to road	
	Distance to stream	
	Population density	
	Soil: ACL	
	Soil: ACSL	
	Soil MSU	

Demand Scenario

Scenario 1: Baseline Data

In this scenario, the land use allocations are based on the continuous land use conversions for the past decades. The rate of land use change is computed for brushwood, grassland, and forest through data projection and linear regression using years 1978 and 1999 as baseline data. Cultivated ISF areas, on the other hand, are computed as a function of population, the most influential factor in land use studies. Both alienable and disposable and lake land uses are considered constant. Also, demand for each land use type is derived independently.

Results for baseline scenario show a high rate of reforestation for the watershed since computation of data is based on linear regression. An increase in cultivated ISF is also seen in some areas where settlements are located. It can be seen that there are no changes for the alienable and disposable land cover type because conversion from this land use type is not allowed.

6.2. Drivers of Land Use Change

6.2.1. Carbon Budgets of Terrestrial Ecosystems in PCW

There is great interest on the role of terrestrial ecosystems in the global carbon cycle. It is estimated that about 60 Pg C is exchanged between terrestrial ecosystems and the atmosphere every year, with a net terrestrial uptake of 0.7 ±1.0 Pg C (Schimmel et al. 1995).

Table 6.3. Conversion matrix

	AandD	B	C	F	G	L	R
AandD / paddy	+	-	-	-	-	-	-
Brushwood	-	+	+	+	+	-	+
Cultivated ISF	-	+	+	+	+	-	+
Forest	-	+	+	+	+	-	-
Grassland	-	+	+	+	+	-	+
Lake	-	-	-	-	-	+	-
Reforestation	-	+	+	-	+	-	+

Legend: (AandD) AandD/paddy; (B) brushland; (C) cultivated ISF; (F) forest; (G) grassland; (L) lake; (R) reforestation; (+) conversion possible; (-) conversion impossible.

The world's tropical forests which cover 17.6 M km2 contain 428 Pg C in vegetation and soils (Watson et al. 2000). However, land use change and forestry (LUCF) activities, mainly in the form tropical deforestation, are significant net sources of CO2 that account for 1.6 Pg yr-1 out of the total anthropogenic emissions of 6.3 Pg yr-1 (Houghton et al. 1997, Watson et al. 2000).

Philippine forest ecosystems have likewise been a source and sink of carbon. From the 1500s to the modern era, it is estimated that deforestation has contributed 3.7 Gt C to the atmosphere (Lasco, 1998). Of this amount, 70% (2.6 Gt) is released this century alone. However, present land use cover also absorbs carbon through regenerating forests and planted trees. The vast areas of degraded land in the Philippines in fact offer great potential for carbon sequestration through rehabilitation activities such as reforestation and agroforestry.

In the last five years, several studies have investigated the carbon stocks of forest ecosystems and other land cover types in the Philippines (e.g. Lasco et al. 2001, Lasco et al. 2000). However, the carbon stocks in PCW have not been adequately characterized. Thus, the main objective of the study is to quantify the carbon stocks of the various land cover types in the Pantabangan watershed. Specifically, the study aimed to:

- Determine the biomass and carbon density of the forest ecosystems and other land cover types in PCW;
- Simulate the carbon stocks of natural forest ecosystems using the CO2Fix model; and
- Assess the capacity of the watershed.

The field methods and the CO2Fix model are described earlier in Section 2.2.2.

6.2.2. Above-Ground Biomass and Carbon Density

Field measurements coupled with the use of allometric equation showed a wide range of biomass density in the various land cover in PCW (Table 6.4 and 6.5). Two estimates are presented here, one using the allometric equation from Brown (1997) and the other using Power Fit developed at the Environmental Forestry Programme (Banaticla 2002). It will be noted that Brown's equation gave about 50% higher biomass estimate than the Power fit equation which is also the same trend in other Philippines studies (Banaticla 2002). Without site-specific biomass equations for the PCW, these two estimates can be used as a high and low estimate for the biomass in the watershed.

Table 6.4. Biomass of various land uses in PCW using allometric equation of Brown (1997)

Land use	Biomass density (mg ha-1)			
	Tree	Herba-ceous/Understorey	Litter	TOTAL
Second Growth	546.6	0.16	16.53	563.29
Brushland	113.35	0.42	7.57	121.34
Grassland		17.15		17.15
Acacia auriculiformis	286.01	0.04	0.62	286.67
Benguet Pine	181.22	0.42	0.83	182.47
Eucalyptus	54.29	0.22	0.72	55.23
Gmelina	108.97	0.15	0.66	109.78
Ipil-ipil	82.55	0.24	0.39	83.18
Mahogany	152.56	0.08	0.41	153.05
Mixed species	87.19	0.4	2.99	90.58
Narra	145.24	0.33	0.72	146.29

Among all the land cover types, secondary forests have the highest above-ground biomass density while grasslands have the lowest. Secondary forests form the only remaining natural forest cover in the watershed. The original forest cover has been slowly decimated over the years as result of timber cutting and shifting cultivation activities. After years of repeated cultivation and burning, grasslands become the dominant vegetative cover.

In attempt to revegetate the denuded areas, there have been several attempts to reforest the watershed. The various reforestation species have biomass density values (about 70 Mg on the average using Power Fit equation) that are lower than natural forests but much higher than grassland areas. This implies that reforestation activities are helping increase the biomass of denuded areas. However, their biomass is typically lower than natural forests.

Table 6.5. Biomass of various land uses in PCW using the Power Fit equation

Land use	Biomass density (Mg ha-1)			
	Tree	Herbaceous/Understorey	Litter	TOTAL
Second Growth	282.91	0.16	16.53	299.6
Brushland	72.38	0.42	7.57	80.37
Grassland		17.15		17.15
Acacia auriculiformis	120	0.04	0.62	120.66
Benguet Pine	95.67	0.42	0.83	96.92
Eucalyptus	28.47	0.22	0.72	29.41
Gmelina	55.66	0.15	0.66	56.47
Ipil-ipil	42.95	0.24	0.39	43.58
	Tree	Herbaceous/Understorey	Litter	TOTAL
Mahogany	80.43	0.08	0.41	80.92
Mixed species	45.38	0.4	2.99	48.77
Narra	76.09	0.33	0.72	77.14
Ave for tree plantations	68.08	0.24	0.92	69.23

The results of the study are consistent with above-ground biomass density values obtained in other parts of the country using the Brown (1997) equation (Lasco and Pulhin 2003, Lasco et al. 2002, Lasco et al. 2000, Kawahara et al. 1981). For example, a secondary forest in Makiling Forest is found to have 547 Mg ha-1 (Lasco et al. 2001) while in Leyte a similar forest type has 446 Mg ha-1 (Lasco et al. 2002). The results of the study are also consistent with the IPCC default values for natural forests in the Philippines which is 370-520 Mg ha-1 (Houghton et al. 1997). Similarly, the biomass density is within the range of other forest types in Southeast Asia (Lasco 2002).

It will be noted that except for grassland areas, most of the above-ground biomass are stored in trees. This is consistent with findings from other studies where more than 90% of biomass is commonly found in the bigger trees (Lasco et al. 1999, Gillespie et al. 1992).

The plantation species have differing biomass density. This could be due to a number of reasons such as age differences between and among species. In addition, it could also be due to the uneven site conditions found in the PCW.

As expected, the carbon density values of the various land cover follow the trend of biomass density (Table 6. 4). Similarly, the area distribution of carbon density is a reflection of the land cover types of the watershed (Figure 6.4).

Allowed Land Use Transition Sequences

Land transition in the PCW is illustrated in Figure 6.2. A forest can be converted to grassland due to deforestation or logging. Brushwood can be converted to grassland due to deforestation. Cultivated ISF or reforestation can be developed from grassland. Forest can remain as forest as long as conservation practices are performed. The conversion matrix (Table 6.3) indicates which land use type is allowed to be converted to other land use type.

On the basis of the carbon density values, the entire watershed is estimated to contain 4,878 Gg C (Power Fit) to 8,870 Gg C (Brown) in above ground biomass and necromass. As expected, most of these are contained in the natural forests (Figure 6.4).

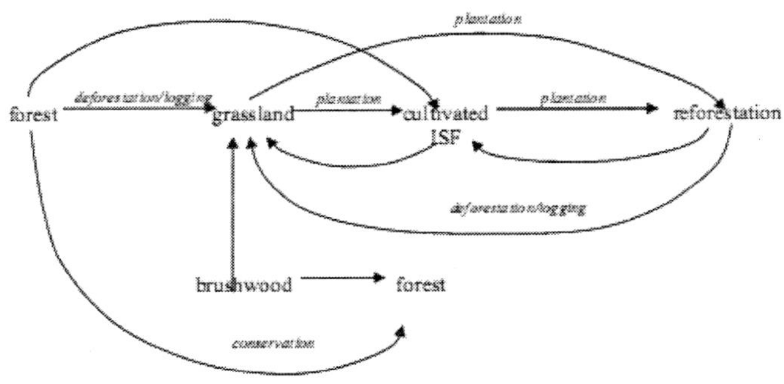

Figure 6.2. Possible land use transitions in PCW.

6.2.3. Simulation of Carbon Budgets Using CO2Fix

Three modules in the CO2 Fix v.2.0 are parameterized using local data where available, viz., general, biomass and soil parameters. Since there is no harvesting, the products module is not used.

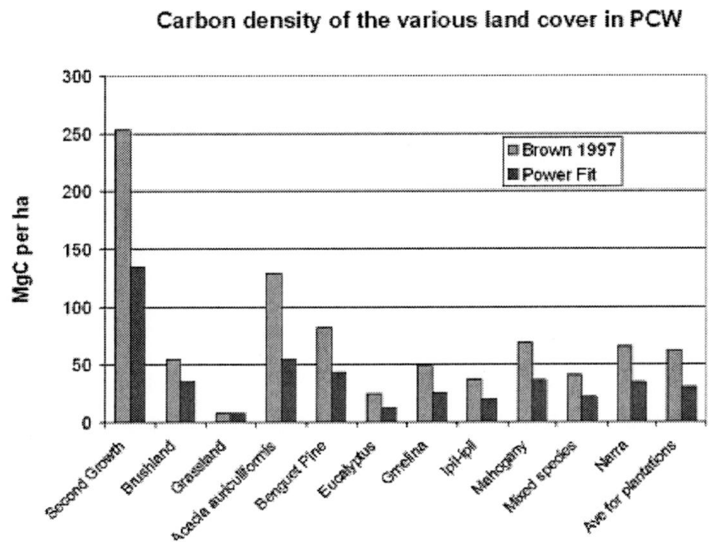

Figure 6.3. Carbon density of land cover types in PCW (Note: carbon content of biomass= 44%).

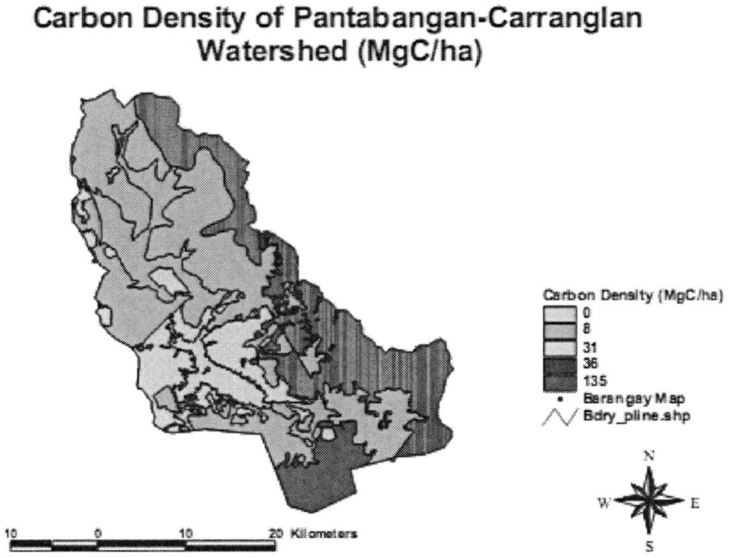

Figure 6.4. Distribution of carbon density in PCW.

The following are the values input to the model:

General Parameters

- Simulation length= 100 yr
- Maximum biomass= 400 Mg (roughly the average of the high and low estimate in Tables 1 and 2)
- Cohorts: Upper storey, understorey
- Cohorts age= 20 years

Biomass Parameters

1. Upperstorey:

- Carbon content= 45% (from Lasco and Pulhin 2003)
- Wood density= 0.57 (Brown 1997)
- Initial carbon= 100 MgC ha-1
- Stem growth rate:

The stem growth rate of the forest is estimated using the following logistic equation of a dipterocarp plantation in the Philippines:

Logistic Model:

y=a/(1+b*exp(cx))

where:

a = 260.9623

b = 204.2171

c = 0.14702

r = 0.998

R2= 0.996

Note: Equation generated using Curve Expert 1.3
The estimated total biomass and CAI are shown in Table 6.6 and 6.7.

- Default values are used for foliage, branches and roots
- Mortality and competition are assumed to be zero.

2. Understorey:

- Carbon content= 45%
- Wood density= 0.57
- Initial carbon= 10 MgC ha-1
- Default values for potential species are used in the other parameters.

Soil Parameters

- Annual mean temperature= 27.6 0C for Cabanatuan station (www.worldclimate.com)
- Precipitation in growing season = 2,000 mm
- PET in growing season= 1,848 from (mean monthly temperature from www.worldclimate.com)

1. 1. Calibration of initial soil carbon

Total litter fall= 10 t ha-1 yr-1 (8 upper; 2 under)
Upper (3 leaves; 2stems; 1 branch; 2 roots)
Under (0.5 leaves; 0.5 stems; 0.5 branch; 0.5 roots)

The results of simulation showed that while carbon in forest biomass is increasing over time by about 50 Mg C per century in the PCW, the soil organic carbon is declining by roughly similar amount. Thus, overall, the total carbon density remains stable over time after an initial decrease.

6.2.4. Potential for Carbon Sequestration

Open areas in the watershed, mainly grasslands and brushwood, have some potential to sequester carbon through tree planting and agroforestry. Because of the harsh and sub-marginal conditions in grassland areas, the rate of carbon sequestration is estimated to be generally less than 4 Mg ha-1 yr-1 (Table 6.6). This is low compared to the IPCC default for tropical plantations (IPCC 1996) as well previous findings in other part of the country like the Makiling Forest Reserve and Leyte where conditions are much better (Lasco and Pulhin 2003, Lasco et al. 2002, Lasco et al. 2001). However, given the large open area in the watershed, there is some potential for carbon sequestration through natural and/or artificial regeneration.

In summary, the study shows that natural forests have a carbon density of 300 and 563 Mg C ha-1 in aboveground biomass and necromass using the Powerfit equation and Brown (1997) equation, respectively. Brushlands and tree plantations have lower carbon densities (generally less than 200 Mg C ha-1) while grasslands have less then 20 Mg C ha-1. Total above-ground carbon stocks of the whole watershed is estimated to range from 4,800 to 8,900 Mg C depending on the biomass allometric equation used. The results of simulation using CO2Fix model showed that while carbon in forest biomass is increasing over time by about 50 Mg C per century in PCW, the soil organic carbon is declining by roughly similar amount. Thus, overall, the total carbon density remains stable over time after an initial decrease. Finally, the watershed has great potential for carbon sequestration through tree establishment in open areas.

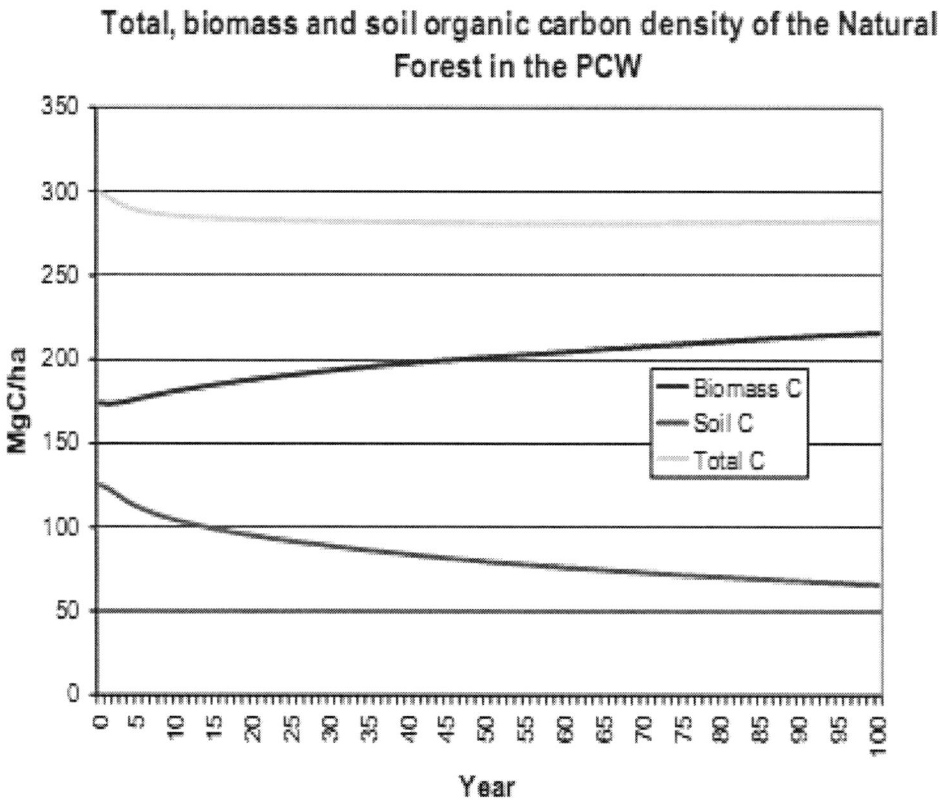

Figure 6.5. Total, biomass and soil carbon density in the natural forest of PCW for 100 years as simulated by CO2Fix.

Table 6.6. Estimated total biomass of a dipterocarp plantation in a good site in the Philippines

Age	Merchantable volume[1]	Biomass	MAI biomass
20	8	5	3
25	60	34	18
30	111	63	10
35	165	94	8
40	223	127	8
45	271	154	6
50	308	176	4
55	330	188	2
60	348	198	3
70[2]		199	0
80		200	0
90		200	0
100		200	0

[1]Merchantable volume data for ages 20 to 60 from PCARDD (1985).
[2]Biomass for ages 70 to 100 are extrapolated from the logistic equation.

Table 6.7. CAI used for stem in the CO2 Fix Model (estimated based on the logistic equation)

Biomass/maximum biomass	CAI (m3 ha-1 yr-1)
0.10	0.7
0.20	3.5
0.30	9.2
0.40	12.2
0.50	12.7
0.60	10.8
0.70	7.6
0.80	4.5
0.90	1.6
1.00	0.5

Table 6.8. Biomass and C density and MAI in Nueva Ecija, Philippines (from Lasco, 2001)

Species	Age (yr)	Ave dbh (cm)	Biomass (Mg ha-1)	MAI (Mg ha-1 yr-1)	C density (Mg ha-1)	MAI (Mg ha-1 yr-1)
Acacia auriculiformis 1	6	5.68	7.39	1.23	3.33	0.55
A. auriculiformis 2	6	6.46	9.97	1.66	4.49	0.75
A. auriculiformis 3	9	9.62	42.51	4.72	19.13	2.13
A. auriculiformis 4	9	8.71	32	3.56	14.4	1.6
A. auriculiformis 5	9	10.47	46.11	5.12	20.75	2.31
A. auriculiformis 6	9	8.73	39.73	4.41	17.88	1.99
Tectona grandis 1	13	5.5	8.7	0.67	3.92	0.3
T. grandis 2	13	7.36	22.3	1.72	10.04	0.77
Gmelina arborea 1	6	7.33	17.22	2.87	7.75	1.29
G. arborea 2	6	6.8	7.71	1.29	3.47	0.58
Pinus kesiya	13	12.53	107.83	8.29	48.52	3.73
P. kesiya + broadleaf spp.	13	10.1	83.24	6.4	37.46	2.88

References

Achard F, Eva HD, Stibig HJ, Mayaux P, Gallego J, Richards T, Malingreau JP. 2002. Determination of deforestation rates of the world's humid tropical forests. Science 297: 999-1002.

Adger WN, Arnell NW, Tompkins EL. 2005. Successful adaptation to climate change across scales. Global Environmental Change 15(2005):77-86.

Antle J, Stoorvogel J, Bowen W, Crissman C, Yanggene D. 2003. The tradeoff analysis approach: lessons from Ecuador and Peru. Quarterly Journal of International Agriculture 42(2): 189–206.

Banaticla MRN, Sales RF, Lasco RD. 2007. Biomass equations for tropical tree plantation species in young stands using secondary data from the Philippines. Annals of Tropical Research 29(3):73–90.

Brady NC. 1984. The nature and properties of soil, 9th ed. New York: Macmillan Publishing Co., Inc.

Bantayan NC, Saplaco SR, Cruz RVO. 2001. GIS-based atlas of selected watersheds in the Philippines. DOST-PCARRD and UPLB-CFNR-ERSG.

Brooks N, Adger N. 2004. Assessing and enhancing adaptive capacity. Technical Paper No 7. In: Adaptation Policy Framework. UNDP.

Brooks N. 2003. Vulnerability, risk and adaptation: a conceptual framework. Tyndall Centre for Climate Change Research Working Paper 38.

Brown K, Adger WN, Tompkins E, Bacon P, Shim D, Young K. 2000. Trade-off analysis for marine protected area management. CSERGE Working Paper GEC 2000-02, Centre for Social and Economic Research on the Global Environment University of East Anglia Norwich, UK.

Brown S. 1997. Estimating biomass and biomass change of tropical forest: A primer. Forestry Paper 134. Food and Agriculture Organization: Rome.

Brown S, Sathaye J, Cannel M, Kauppi P. 1996. Management of forests for mitigation of greenhouse gas emissions. In: Watson RT, Zinyowera MC, Moss RH, eds. Climate Change 1995: Impacts, Adaptations, and Mitigation of Climate Change: Scientific-Technical Analyses, Contribution of Working Group II to the Second Assessment Report of the Intergovernmental Panel on Climate Change. Cambridge and New York, Cambridge University Press, 775-797.

Burton I. 1998. Climate adaptation policies for Canada. Option Politiqiues. p.8-9.

Chua LA. 1999. Understanding the research process. 3rd ed. Department of Agricultural Education and Rural Studies, College of Agriculture, UPLB.

Cruz RVO, Lasco RD, Pulhin JM, Pulhin FB, Garcia KB. 2004. Assessment of climate change impacts, vulnerability and adaptation: water resources of Pantabangan-Carranglan Watershed. AIACC Working Paper No 9. Environmental Forestry Programme, College of Forestry and Natural Resources, UPLB.

Cruz RVO, Saplaco SR, Lasco R D, Avanzado MMB, Pulhin FB. 2000. Development of water budget models for selected watersheds in the Philippines: Assessment of the impacts of ENSO on the water budget of selected watersheds. Unpublished research report. Philippine Council for Agriculture, Forestry and Natural Resources Research and Development. Department of Science and Technology.

Cruz RVO. 2002. Climate change and water resources: impacts, adaptation and vulnerability assessment. Lecture presented during the Training-Workshop on Research Methods in Assessing Climate Change Impacts, Adaptation and Vulnerability in Watershed Areas and Communities held on 25 November–7 December 2002 at the Molave Training Room, TREES, CFNR, UPLB, College, Laguna.

Cruz RVO. 1997. Adaptation and mitigation measures for climate change: impacts on the forestry sector. In: Proc. Consultation Meeting for the International Conference on Tropical Forests and Climate Change. Environmental Forestry Programme, College of Forestry and Natural Resources, UPLB.

Daño AM. 1994. Effects of land use change on soil and hydrologic characteristics of experimental watersheds in Angat, Bulacan. PhD in Soil Science Dissertation. UPLB.

[DENR-UNEP] Department of Environment and Natural Resources-United Nations Environment Programme. 1997. Philippine biodiversity: an assessment and action plan. Makati City: Bookmark Inc.

[DMPMC] Development Master Plan of the Muncipality of Carranglan, Nueva Ecija 2003-2007.

Doikno GR. 1997. Coping with El Niño of the century. Canopy 23(5):1, 6-9.

Epstein ES. 1991. On obtaining daily climatological values from monthly means. J. Climate 4:365-368.

Esman MJ, Bruhns E. 1965. Institution building in national development. an approach to induced social change in transitional societies. Graduate School of Public and International Affairs, University of Pittsburgh.

Federer, C A. 1995. Brook 90: A simulation model for evaporation, soil water and streamflow, version 3.1 computer free ware and documentation. USDA Forest Service.

[FMB] Forest Management Bureau. 1998. Forestry statistics (1997). Forest Management Bureau, Quezon City, Philippines.

Forest Development Center. 1987. Towards a successful national reforestation program. Policy Paper No. 24. College of Forestry and Natural Resources, UPLB.

Garrity DP, Kummer DM, Guiang ES. 1999. The upland ecosystem in the Philippines: alternatives for sustainable farming and forestry. Washington, DC: Natioal Academy Press.

Gillespie AJR, Brown S, Lugo AE. 1992. Tropical forest biomass estimationa from truncated stand tables. For Ecolo and Mgt 48: 69-87.

Geoghegan J, Cortina Villar S, Klepeis P, Macario Mendoza P, Ogneva-Himmelberger Y, Roy Chowdhury R, Turner II BL, Vance C., 2001: Modeling tropical deforestation in the southern Yucata´ n peninsular region: comparing survey and satellite data. Agriculture, Ecosystems, and Environment 85: 25–46.

Goldstein H. 1995. Multilevel statistical models, 2nd ed. London: Arnold and New York: John Wiley and Sons.

Hewitson BC. 2003. Developing perturbations for climate change impact assessments. EOS 84(35):337-348.

Holdridge LR. 1967. Life Zone Ecology. revised ed. San Jose, Costa Rica: Tropical Science Center.

Houghton JT, Meira Filho LG, Lim B, Treanton K, Mamaty I, Bonduki Y, Griggs DJ, Callander BA, eds. 1997. Greenhouse Gas Inventory Workbook. Intergovernmental Panel on Climate Change (IPCC), Organization for Economic Co-operation and Development (OECD) and the International Energy Agency (IEA), Paris.

[IPCC] Intergovernmental Panel on Climate Change. 1996. Climate change 1995. The science of climate change. Contribution of the Working Group I to the Second Assessment Report of the Intergovernmental Panel on Climate Change. World Meteorological Organization, Geneva, Switzerland and United Nations Environment Programme, Nairobi, Kenya.

IPCC] Intergovernmental Panel on Climate Change. 2001. Climate change 2001: impacts, adaptation, and vulnerability. Contribution of Working Group II to the Third Assessment Report of the Intergovernmental Panel on Climate Change. Cambridge: Cambridge University Press.

[IPCC] Intergovernmental Panel on Climate Change. 2001. The regional impacts of climate change:. an assessment of vulnerability. A special report of the IPCC Working Group II. Cambridge: Cambridge University Press.

[IPCC] Intergovernmental Panel on Climate Change. 2003. Good practice guidance for land use, land-use change and forestry. Penman J, Gytarsky M, Hiraishi T, Krug T, Kruger D, Pipatti R, Buendia L, Miwa K, Ngara T, Tanabe K, Wagner F, eds. Kanagawa, Japan.

[IPCC] Intergovernmental Panel on Climate Change. 2006. 2006 IPCC guidelines for national greenhouse gas inventories. Prepared by the National Greenhouse Gas Inventories Programme. Eggleston HS, Buendia L, Miwa K, Ngara T, Tanabe K, eds. IGES, Japan.

[IPCC] Intergovernmental Panel on Climate Change. 2007. Climate change 2007: impacts, adaptation and vulnerability. Contribution of Working Group II to the Fourth Assessment Report of the Intergovernmental Panel on Climate Change. Parry ML, Canziani OF, Palutikof JP, van der Linden PJ, Hanson CE, eds. Cambridge: Cambridge University Press.

Jones R, Boer R. 2003. Assessing current climate risks. Technical Paper No. 5. Adaptation Policy Framework. New York: UNDP.

Jose AM, Sosa LM, Cruz NA. 1996. Vulnerability assessment of Angat Water Reservoir to climate change. Water, Air and Soil Pollution 92: 191-201.

Kanninen M, Murdiyarso D, Seymour F, Angelsen A, Wunder S, German L. 2007. Do trees grow on money? The implications of deforestation research for policies to promote REDD. Center for International Forestry Research (CIFOR), Bogor, Indonesia.

Kawahara T, Kanazawa Y, Sakurai S. 1981. Biomass and net production of man-made forests in the Philippines. J Jap For Soc 63(9): 320-327.

Kummer DM. 1990. Deforestation in the post-war Philippines. Ph.D. dissertation, Boston, Boston University.

Lasco RD, Lales JS, Arnuevo MT, Guillermo IQ, de Jesus AC, Medrano R, Bajar OF, Mendoza CV. 2002. Carbon dioxide (CO2) storage and sequstration of land cover in the Leyte Geothermal Reservation. Renewable Energy 25:307-315.

Lasco RD, Lales JS, Guillermo IQ, Arnouevo T. 1999. CO2 absorption study of the Leyte geothermal forest reserve. Final report of a study conducted for the Philippine National Oil Company. UPLB Foundation Inc.

Lasco RD, Pulhin FB, Visco RG, Racelis DA, Guillermo IQ, Sales RF. 2000. Carbon stocks assessment of Philippine forest ecosystems. Paper presented at the Science-Policy Workshop on Terrestrial Carbon Assessment for Possible Carbon Trading. Bogor, Indonesia. 28-20 February 2000.

Lasco RD, Pulhin FB. 1998. Philippine forestry and CO2 sequestration: opportunities for mitigating climate change. Environmental Forestry Programme, College of Forestry and Natural Resources, UPLB.

Lasco RD, Pulhin FB. 2003. Philippine forest ecosystems and climate change: carbon stocks, rate of sequestration and the Kyoto Protocol. Annals of Tropical Research 25(2):37-51.

Lasco RD. 1998. Management of tropical forests in the Philippines: implications to global warming. World Resource Review 10(3): 410-418.

Lasco RD. 2002. Forest carbon budgets in Southeast Asia following harvesting and land cover change. Science in China 45:55-64.

Liang X, Lettenmaier DP, Wood EF, Burges SJ. 1994. A simple hydrologically based model of land surface water and energy fluxes for GSM. J. Geophysics Res. 99(97):415-428.

Liu DS, Iverson LR, Brown S. 1993. Rates and patterns of deforestation in the Philippines: application of geographic information system analysis. Forest Ecology and Management 57, 1-16.

Lohmann D, Raschke E, Nijssen B, Lettenmaier SJ. 1998. Regional scale hydrology: Formulation of the VIC-2L model coupled to a routing model. Hydrological Sciences Journal 43(1):131-141.

Mayaux P, Holmgren P, Achard F, Hugh E, Stibig HJ, Branthomme A. 2005. Tropical forest cover change in the 1990s and options for future monitoring. Philosophical Transactions of the Royal Society B: Biological Sciences 360(1454): 373-384.

McCaskill MR. 1990. An efficient method for generation of full climatological records from daily rainfall. Aust. J. Agric. Res. 41:595-602.

McNeely JA, Miller KR, Reed WV, Mitternmeier RA, Werner TB. 1990. Conserving the world's biological diversity. Washington, DC: IUCN, WRI, CI, WWF-US and the World Bank.

[MPMP] Master Plan of the Municipality of Pantabangan, Nueva Ecija 1998-2000.

[NDCC] National Disaster Coordinating Council. 2006. Media update, Situational Report No. 29 Effects of typhoon "MILENYO" (Xangsane), October 14, 2006, 8:00am. National Disaster Coordination Center, Manila.

[NEDA] National Economic Development Authority. 2002. National framework for physical planning 2001-2030. National Land Use Committee, National Economic Development Authority. Pasig City, Philippines.

[NIA] National Irrigation Administration. [no date]. Dam and Reservoir Division. Upper Pampanga River Integrated Irrigation System.

Niang-Diop I, Bosch H. 2004. Formulating an adaptation strategy. Technical Paper No 8. In: Adaptation Policy Framework. UNDP.

[NIA-UPRIIS] National Irrigation Authority-Upper Pampanga River Integrated Irrigation System. 2004. UPRIIS Firmed-up Service Area (in hectares by municipality). Based on parcellary mapping.

[NPC] National Power Corporation. 1995. Pantabangan Watershed Rehabilitation Project. Watershed Management Department. Quezon City.

[NPC] National Power Corporation. 1997. Pantabangan-Carranglan Watershed Management Plan. Watershed Management Department. Quezon City.

[NSCB] National Statistical Coordination Board. 2000. Philippine statistical yearbook. Makati City.

PAGASA-DOST. 2001. Drought episodes in the Philippines. Climate Data Section, Climatology and Agrometeorology Branch, Quezon City, Philippines

PAGASA-DOST. 1997. Primer on El Niño/Southern Oscillation (ENSO). Climatology and Agrometeorology Branch, Quezon City, Philippines.Pulhin JM, Peras RJJ, Cruz RVO, Lasco RD, Pulhin FB, Tapia MA. 2004. Vulnerability of watershed communities to climate variability and extremes in the Philippines. AIACC Working Paper No 8. Environmental Forestry Programme, College of Forestry and Natural Resources, UPLB.

Pulhin JM. 2002. Climate change and watershed communities: methodology for assessing social impacts, vulnerability and adaptation. A paper discussed during the AIACC - AS 21 Regional Capability-Building Training-Workshop on Climate Change Impacts, Adaptation and Vulnerability, College of Forestry and Natural Resources, UPLB. 25 November–8 December 2002.

Pearson TS, Walker S, Brown S. 2005. Sourcebook for land use, land use change and forestry projects. Winrock International.

Polsky C, Easterling III WE. 2001. Adaptation to climate variability and change in the US Great Plains: a multi-scale analysis of Ricardian climate sensitivities. Agric. Ecosyst. Environ. 85, 134-144.

Pontius RG, Schneider LC. 2001. Land use change model validation by an ROC method for the Ipswich watershed, Massachusetts, USA. Agriculture, Ecosystems and Environment. 85:239-248.

Revilla AV. 1997. Working paper for the forestry policy agenda for the incoming administration, College of Forestry and Natural Resources, UPLB.

Saplaco SR, Bantayan NC, Cruz RVO. 2001. GIS-based atlas of selected watersheds in the Philippines. DOST-PCARRD and UPLB-CFNR-ERSG.

Schimmel D, Enting IG, Heimann M, Wigley TML, Rayneud D, Alves D, Seigenthler U. 1995. CO2 and the carbon cycle. In: Houghton JT, Meira Filho LG, Bruce J, Lee H, Callander BA, Haites E, Harris N, Maskell K, eds. Climate change 1994: radiative forcing of climate change and an evaluation of the IPCC IS92 emision scenarios. Cambridge: Cambridge University Press. p 35-71.

SEA START-RC. 2002. SEA-BASINS Integrated Hydrological Modeling System.

Serneels S, Lambin EF. 2001. Impact of land-use changes on the wildebeest (Connochaetes taurinus) of the northern part of the Serengeti-Mara ecosystem. Journal of Biogeography 28(3), 391-408.

Somaratne S, Dhanapala AH. 1996. Potential impact of global climate change on forest distribution in Sri Lanka. Water, Air, and Soil Pollution 92:129–135.

Stern RD, Coe R. 1984. A model fitting analysis of daily rainfall data. J.R. Statist. Soc. A 147:1-37.

Stoorvogel JJ, Antle JM, Crissman CC, Bowen W. 2004a. The tradeoff analysis model: integrated bio-physical and economic modeling of agricultural production systems. Agricultural Systems 80: 43–66.

Stoorvogel JJ, Antle JM, Crissman CC. 2004b. Trade-off analysis in the Northern Andes to study the dynamics in agricultural land use. J of Environmental Management 72:23–33.

Swets JA. 1986: Form of empirical ROCs in discrimination and diagnostic tasks: Implications for theory and measurement of performance. Psychol. Bull.,99, 181–198.

[PINC] Philippines' Initial National Communication on Climate Change. 1999. Republic of the Philippines.

Toquero FD. 2003. Impact of involuntary resettlement: the case of Pantabangan Resettlement in the Province of Nueva Ecija. Unpublished PhD thesis. Central Luzon State University, Science City of Muñoz, Nueva Ecija.

Toquero FD. personal communication. 10 January 2005.

[UNEP] United Nations Environment Programme. 2001. Vulnerability indices: climate change impacts and adaptations. Downing, TE, Butterfield R, Cohen S, Huq S, Moss R, Rhaman A, Sokona Y, Stephen L, eds. United Nations Environment Programme Policy Series.

Verburg P. 2002. CLUE-S manual: model version for relatively small areas and data-sets with a high spatial resolution. Wagenigen University, The Netherlands.

Verburg PH, Veldkamp A. 2004. Projecting landuse transitions and forest fringes in the Philippines at two spatial scales. Landscape Ecology 19: 77–98.

Watson RT, Noble IR, Bolin B, Ravindranath NH, Verado DJ, Dokken DJ, eds. 2000. Land use, land-use change, and forestry. Cambridge: Cambridge University Press.

Yoshida S. 2000. 'Pantabangan forestry development assistance project', www.jica.go.jp/english/news/2000/16.html.

Web References

http://www.hydro.washington.edu/Lettenmaier/Models/VIC/VIChome.html.

http://www.worldclimate.com.

Data on Socio-Economic Scenario:

http://sres.ciesin.columbia.edu/tgcia/.

http://sedac.ciesin.columbia.edu/plue/gpw/index.html?main.htmland2.

http://www.ornl.gov/gist/landscan/index.html.

http://www.ciesin.columbia.edu/indicators/ESI).

http://www.census.gov.ph/census2000/index.html/.

Data on Climate Scenario:

http://ipcc-ddc.cru.uea.ac.uk/dkrz/dkrz_index.html.
http://ipcc-ddc.cru.uea.ac.uk/asres/scenario_home.html.
http://ipcc-ddc.cru.uea.ac.uk/asres/sres_home.html.
http://www.start.org/Projects/AIACC_Project/resources.

INDEX

A

B

C

D

E

F